RAVELING THE BRAIN

NEW DIRECTIONS IN RHETORIC AND MATERIALITY

BARBARA A. BIESECKER, WENDY S. HESFORD,
AND CHRISTA TESTON, SERIES EDITORS

RAVELING THE BRAIN

TOWARD A TRANSDISCIPLINARY NEURORHETORIC

JORDYNN JACK

THE OHIO STATE UNIVERSITY PRESS

COLUMBUS

Library of Congress Cataloging-in-Publication Data available online at catalog.loc.gov.

Cover design by Susan Zucker
Text design by Juliet Williams
Type set in Adobe Minion Pro

CONTENTS

ACKNOWLEDGMENTS

THIS BOOK is about raveling the brain. Here, though, I'd like to thank those who kept my brain from *unraveling* as I was writing it.

At the University of North Carolina, Chapel Hill, I received research leave from the Institute for the Arts and Humanities (IAH) and the College of Arts and Sciences. At the IAH, I also enjoyed the intellectual support of my cohort from spring 2012, who read an early proposal and chapter for this manuscript and provided valuable suggestions. I would also like to thank department chairs Beverly Taylor and Mary Floyd-Wilson for supporting my research.

The rhetoric and composition community at UNC continues to stimulate my thinking, including Jane Danielwiecz, Todd Taylor, Daniel Anderson, and Courtney Rivard. The HHIVE Lab, especially Jane Thrailkill, Kym Weed, and my English 695 students, have been listening to me ramble about neuroscience, as did students in my English 805 graduate class in spring 2017, including Oren Abeles, Tyler Easterbrook, Jennifer Edwell, Tiffany Friedman, Sarah Singer, and Carly Wetzel. Research assistants Jennifer Edwell, Dylan Thompson, Erik Mahoney, and Colin Dekeersgieter helped with literature reviews and bibliographic work.

Within the discipline, my colleagues in the Association for the Rhetoric of Science, Technology, and Medicine and the Rhetoric Society of America have been indispensable as a sounding board for my ideas. I would like to thank, especially, David Gruber, Jeremiah Dyehouse, Lynda Walsh, Marika

Seigel, Jenell Johnson, Melissa Littlefield, Jodi Nicotra, Davi Johnson Thornton, James Wynn, Daniel Gross, Jeanne Fahnestock, Carolyn Miller, and John P. Jackson. I also received advice, suggestions, and ideas from Rebecca Dingo, Scott Wible, Jessica Enoch, Debra Hawhee, Melanie Yergeau, and many, many others. I am also indebted to the participants in the Neurorhetorics workshop at the RSA Institute in Minneapolis in 2015. My mentors, Cheryl Glenn, Jack Selzer, and Stuart Selber, continue to show me the way.

My work on this project began in part through conversations with my neuroscience colleagues at Duke University, especially L. Gregory Appelbaum, Scott Huettel, and Elizabeth Beam. They pointed me to key conversations in the discipline about neurorealism, neuroessentialism, and statistical significance, and explained to me the significance of how neuroscience concepts are operationalized in fMRI research. Our collaborations prompted me to think more deeply about how humanistic researchers can work together with neuroscientists.

I've been lucky to have the enthusiastic support of Tara Cyphers and the editorial team at The Ohio State University Press. Two anonymous reviewers provided detailed suggestions that strengthened this book considerably.

Portions of this work have appeared previously in print. Parts of chapter 2, "Neuroessentialism," appear in "How to Do Neurorhetorics: A Tutorial," published in *Itinerations* in February 2013. Parts of chapter 4, "Neurosex," appear in "How Good Brain Science Gets That Way," published in *Feminist Rhetorical Science Studies: Human Bodies, Posthuman Worlds,* edited by Amanda K. Booher and Julie Jung, 164–82 (Carbondale: Southern Illinois University Press, 2018) and in "Hard Core Rhetoric: Gender, Genre, and the Image in Neuroscience," published in *Sexual Rhetorics,* edited by Jacqueline Rhodes and Jonathan Alexander, 58–71 (New York: Routledge, 2015).

Finally, I am grateful for my husband, Ryon, and my children, Penelope and Franklin, both of whom were born while I was working on this project. Without them I would probably have finished this book much earlier, while having much less fun.

INTRODUCTION

I felt a Cleaving in my Mind—
As if my Brain had split—
I tried to match it—Seam by Seam—
But could not make them fit.

The thought behind, I strove to join
Unto the thought before—
But Sequence ravelled out of Sound
Like Balls—upon a Floor.[1]

WHAT DOES it mean to ravel the brain? It is not a common turn of phrase, but it is quite common to speak of "unraveling the brain." Frequently, popular news articles tell us that scientists are now close to "unraveling" the brain's secrets. The same phrase appears in quite a few scientific articles. Perhaps for scientists it is nothing more than a catchy title, a way of jazzing up "Single-Cell RNA Sequencing" or the "Functional Connectivity of Encoding and Retrieval."[2] When presented as something to be unraveled, the brain becomes a mysterious, enigmatic, shadowy entity that is nonetheless passively awaiting discovery. Like the earth, the body, and other objects of scientific inquiry, the brain must be opened up, penetrated, rendered open to the all-knowing scientific gaze. Indeed, the latest scientific tools, such as functional magnetic resonance imaging (fMRI), positron emission tomography, and electronic

---

1. Dickinson, "The Lost Thought," in *The Poems of Emily Dickinson*, edited by Thomas H. Johnson, Cambridge, Mass.: The Belknap Press of Harvard University Press, Copyright © 1951, 1955 by the President and Fellows of Harvard College. Copyright © renewed 1979, 1983 by the President and Fellows of Harvard College. Copyright © 1914, 1918, 1919, 1924, 1929, 1930, 1932, 1935, 1937, 1942, by Martha Dickinson Bianchi. Copyright © 1952, 1957, 1958, 1963, 1965, by Mary L. Hampson.

2. To take just a few examples, consider Katsurakawa and Sakai, "Unraveling Brain Network Coding with a Connectivity-Based Classifier"; Ofengeim et al., "Single-Cell RNA Sequencing"; and Panda et al., "Unraveling Brain Functional Connectivity of Encoding and Retrieval in the Context of Education."

brain stimulation, all promise just that—a newer, unmediated (though always, really, mediated) view of the brain and how it really works. As Donna Haraway and other feminist science studies scholars have indicated, this understanding of the brain presents a masculinist understanding of science that takes for granted the naturalness of objectivity guided by the neutral male gaze.[3]

To unravel the brain also promises simplification. A knitted sweater is a complex item, with threads going in different directions and intricately constructed patterns and stitches. But once you locate the end of the thread and tug, that complex intricacy turns into a simple length of yarn. The idea of "unraveling" the brain suggests a fantasy of sorts: the idea that the brain, too, can be simplified from a complicated, tangled network of neurons and synapses to something as uncomplicated as a single strand of thread. Take, for instance, Daniel Margulies's claim that in order to develop "simplified abstraction[s] of brain organization, . . . some insights may first need to be gleaned by unravelling the tapestry thread by thread."[4] Envisioning the brain as a woven tapestry made up of multiple, interconnected networks, Margulies suggests that, like a sweater, that tapestry can be pulled apart or unraveled.

Emily Dickinson's poem uses a similar metaphor in one of many poems she wrote about the brain. Here, the idea of "Sequence" raveling out of "Sound," like "Balls—upon a floor" might evoke the image of a ball of yarn rolling out, its length of thread unraveling in its path. Yet this image does not work to simplify in Dickinson's case. In general, we do not simply *unravel* poems, teasing out meanings that move from complex to simple; we *ravel them,* tying images and ideas together, generating multiple interpretations, puzzling them out. It is this idea of *raveling* the brain that I will explore in this book.

In particular, I will ravel out a problem—a fundamentally rhetorical problem—that is nonetheless of great interest to neuroscientists, a problem that has to do with the persuasiveness of neuroscience images and research findings. Typically, this problem is attributed to popularization, with the idea being that popular news articles, blogs, and magazines present overhyped claims about the brain that readers then uncritically accept. In 2005, Erik Racine, Ofek Bar-Ilan, and Judy Illes published "fMRI in the Public Eye" in *Nature Reviews Neuroscience,* warning against the risks of popularized neuroscience accounts.[5] Racine, Bar-Ilan, and Illes identified three types of problematic claims in popular accounts of neuroscience: neurorealism, neuroessentialism, and neuropolicy. Neurorealist claims are ones that make neural concepts or mental concepts uncritically real by locating them in the brain. Neuroessen-

---

3. On masculinist vision in science, see Haraway, "Situated Knowledges," 581.
4. Margulies, "Unraveling the Complex Tapestry of Association Networks," 241.
5. Racine, Bar-Ilan, and Illes, "fMRI in the Public Eye."

tialist claims reduce complex concepts to measurable identities in the brain. Neuropolicy occurs when arguers claim prematurely that brain studies justify a particular political goal. Racine, Bar-Ilan, and Illes imply that these types of claims, which I will refer to as "neuroclaims," occur mainly in popular journalism and are an effect of oversimplification or misunderstanding.

Similar concerns have been raised about the use of neuroscience images and even of invoking neuroscience or brain studies at all in popular discourse. In 2008 David P. McCabe and Alan D. Castel published a study of the effects of colorful neuroscience images, called "Seeing Is Believing: The Effect of Brain Images on Judgments of Scientific Reasoning."[6] And in the same year, Deena Skolnick Weisberg and her colleagues published a similar study, "The Seductive Allure of Neuroscience Explanations," only this time dealing with descriptions of neuroscience findings rather than images.[7] Together, these studies suggested a worrisome, persuasive effect of neuroscience information (whether visual or verbal) in popular science texts.

In these studies, neuroscientists primarily refer to experimental studies that use neuroimaging techniques, especially functional magnetic resonance imaging (fMRI). While fMRI is not the only research tool neuroscientists use, it is often featured in popular science, especially for articles reporting on social and cognitive neuroscience, a field that seeks to identify correlations between psychological concepts (such as emotion, reason, and attention) and brain regions or networks. This type of research is often picked up in popular news accounts because it purports to explain phenomena that are readily observable in daily life, but it has also been targeted with numerous critiques for its tendency toward hype.[8]

Throughout this book, when I refer to "neuroscience" I refer to social and cognitive neuroscience that uses fMRI. Of course, neuroscience researchers use a wide range of methods to investigate a wide range of phenomena, including signaling on the cellular level, the connections between neurons that make up the nervous system, and the physiology of disease. A rhetoric-inspired take on these topics is surely possible, but social and cognitive neuroscience lend themselves especially well to rhetorical inquiry because they deal with topics that overlap with our own interests, such as emotion, politics, reason, and persuasion.

At this point, the perils of neurohype are commonly invoked in debates over the public impact of neuroscience, its persuasiveness, and the ethics of

---

6. McCabe and Castel, "Seeing Is Believing."

7. Weisberg et al., "The Seductive Allure of Neuroscience Explanations."

8. See, for instance, Legrenzi and Umiltà, *Neuromania,* and Satel and Lilienfeld, *Brainwashed.*

its use in public debates. And it is not just pundits and commentators who have identified these perils, but neuroscientists themselves. Yet, neuroscientists seem unable to address why the claims associated with neurohype emerge and why (and when) they persuade. I will argue here that the problem is not simply a matter of popularization, but that neuroclaims are embedded in scientific articles, and, more deeply, that they are rooted in the practice and discourse of neuroscience research.

Clearly, much overlap exists between what might be considered "popular science" and "science itself." While the boundary is blurry, here I focus primarily on scientific articles published in academic journals as representative of what neuroscientists would consider "science" and popular books, news and magazine articles, blogs, and so on as "popular science." In my collaborations with neuroscientists, they have reported to me that they focus on academic journal articles as the primary source for the latest research in their field; for the most part, they do not report that they read or pay much attention to popular books or articles because they either represent known information or information that has been adapted for nonspecialist audiences. In part because popular science does not usually make up part of their media diet, it is easy for neuroscientists to suggest that those genres should take the blame for neurohype. Of course, they sometimes do oversimplify and overpromise in reports of the latest findings. Researchers in rhetoric such as Davi Johnson Thornton have shown effectively how these claims circulate and to what effect.[9]

In this book, however, I focus on scientific articles, and I aim to show how rhetoric can offer a deeper analysis of how neuroclaims emerge from those texts. While deemed fallacies or sources of bias by neuroscientists, the three neuroclaims (neurorealism, neuroessentialism, and neuropolicy) actually correspond to three fundamental questions upon which knowledge can be built: questions of fact (does it exist?), definition (what is it?), and policy (what do we do about it?). Indeed, I will show that neuroclaims are not fallacies but the very basis for argument within the neurosciences. They form the rhetorical groundwork for scientists seeking to advance knowledge claims about the brain. What is more, these claims do not simply function at the level of discourse or representation but are embedded in the research methods, tools, and techniques used in the neurosciences, and they arise out of a long history of scientific practice. For this reason, they cannot simply be rooted out as sources of bias but must be recognized as part of the scientific apparatus for social and cognitive neuroscience as it is currently practiced.

---

9. Thornton, *Brain Culture.*

Neuroclaims constitute what I am calling the rhetorical-material mesh-work of the discipline. By *rhetorical-material meshwork*, I refer to the set of interconnected discourses, arguments, bodies, tools, techniques, practices, theories, and traditions that neuroscientists use to do research.[10] This rhetori-cal-material meshwork represents the grounds from which research problems are chosen, problems are framed, concepts are understood, and bodies (and their brains) are recruited to produce scientific knowledge. We can under-stand this meshwork better not by unraveling it, but by raveling—by following threads of discourse across time and through their different movements: we will imagine ourselves, so to speak, as the weaver's shuttle moving in and out of these threads.

Scholars of rhetoric recognize the three questions I mentioned above as points of stasis, a concept they use to understand how arguments function. In ancient rhetoric, stasis was "the method by which rhetors . . . identified the area of disagreement, the point that was to be argued, the issue on which a case hinged."[11] According to Jeanne Fahnestock and Marie Secor, the stasis questions constitute "a taxonomy of arguments" that follows a "logical, hier-archical order."[12] The stases are usually listed as follows:

- Fact—does it exist?
- Definition—what is it?
- Evaluation—is it good or bad?
- Policy—what should be done about it?

Typically, the stasis questions are defined as moving from the "lower" ques-tions of fact and definition to the "higher" questions of evaluation and policy. In scientific articles, however, authors tend to rest at the stasis of fact and defi-nition. However, as Fahnestock and Secor suggest, "When we address an issue in one stasis, we may do so because of the preferred practices of our discipline. That does not mean, however, that the other stases are absent or irrelevant to the reactions of our readers or audience."[13] Clearly, issues of evaluation and policy are relevant to social and cognitive neuroscience, especially given the ethical implications of techniques so intimately connected with human minds,

---

10. Following anthropologist Tim Ingold, who writes eloquently about weaving as a way of conceptualizing such connections, I prefer the term *meshwork* to the more common *network* because it better evokes the idea of threads that are crisscrossed, intricately connected into a fabric, as opposed to a set of connected points. See Ingold, *Being Alive*, 63.

11. Carter, "Stasis and Kairos," 98.

12. Fahnestock and Secor, "The Stases in Scientific and Literary Argument," 428.

13. Fahnestock and Secor, 431.

but these issues tend to be taken up in popular science rather than scientific articles.

While thinking through the stasis questions may seem to offer one way to *unravel* a discourse, it is notable that the term *stasis* comes from the same root as the term *histos,* or loom. *Stasis* referred to the "place one stands" (and hence, metaphorically, to position, opinion, statue, or decree); the looms used in ancient Greece were vertical and would have stood upright (or leaned against a wall). The term *histos* came to refer not only to the loom itself but to the woven fabric it produced. This etymology links stasis with another form of raveling, then: weaving. Through the stasis questions, we might not simply *unravel* but *ravel,* weaving together questions of discourse, practices, and tools to gain a deeper, richer understanding of neuroscience. Asking how scientific topics are defined, for instance, leads to a deeper investigation of how fMRI-based neuroscience experiments are conducted, how research determine what "counts" as envy or love or politics. Asking how facts are established in neuroscience leads to an examination of scientific rhetoric, particularly in the scientific article, but also to an examination of the experiment itself as a site of rhetorical practice.

SIGNIFICANCE

I am not the first to engage in neuroscience from a rhetorical perspective: Scholars in rhetoric have actively engaged in such questions for several decades at minimum, as questions about the nature of cognition naturally impinge upon rhetoric. As Jeffrey Walker argues, "Every rhetoric presupposes a psychology, and every psychology presupposes an account of brain function, a psychobiology."[14] One could even argue, given this relationship between rhetoric and psychology, that our discipline's engagement with the mind goes much deeper, to the earliest stages of rhetorical inquiry itself.

One of the earliest (Western) rhetorical texts we have, the *Encomium of Helen,* might be understood in this light—if not as an account of brain function, then at least as an account of the intersections of what we would now call rhetoric and psychology. Famously, Gorgias likens "the effect of speech upon the condition of the soul" to "the power of drugs over the nature of bodies." More specifically, he suggests that "just as different drugs dispel different secretions from the body," so too with speeches, "some distress, others delight, some cause fear, others make the hearers bold, and some drug

---

14. Walker, "Of Brains and Rhetorics," 301.

and bewitch the soul with a kind of evil persuasion."[15] Surely, here we have a
theory of rhetorical psychology (or psychological rhetoric): Gorgias describes
how different types of speeches convey specific emotional effects, and posits a
"condition of the soul" that accounts for those effects.[16] From this description,
it is easy to imagine a modern-day Gorgias taking great interest in neuroscientific studies that purport to reveal those conditions. Of course, Aristotle's
corpus also includes reflections on what might be classified as psychology (to
use the modern term) in *De Anima, De Sensa,* and *De Memoria,* as well as in
the *Art of Rhetoric,* and these works reflect considerations about the materiality of psychological states as well. Interest in these questions persists throughout rhetorical traditions, including the faculty psychology popularized by
Campbell, to Burke's consideration of the psychology of form, to more recent
investigations seeking to link rhetoric to the neuroscience, such as Thomas
Rickert's study of ambient rhetoric.[17]

Recently, the term *neurorhetoric* has been used to describe this area of
inquiry, following a trend across the disciplines where the prefix *neuro-* has
been applied, from neuroethics to neurophilosophy. Typically, the prefix
neuro- signals an attempt to apply neuroscientific findings to the discipline in
question. In rhetoric, however, the use of the prefix *neuro-* refers not only to
scholarship that seeks to apply neuroscientific insights or integrate them into
a discipline, but also to study of the rhetoric of neuroscience itself.

One main strand in neurorhetoric has focused primarily on traditional
understandings of what falls under the purview of rhetoric: arguments, language, texts. For example, Matt Mays and Julie Jung define neurorhetoric in
such a way that it remains primarily a linguistic inquiry into the language,
terminology, and arguments used by those advancing neuroscience claims.
They state that neurorhetoric "takes as its subject discursive networks constitutive of contemporary neuroscience" and argue that "neurorhetoric . . .
consciously attends to the ways in which shared terminology can induce readers to mistake arguments for empirical proofs."[18] While they do not wish to
leave scientific knowledge unexamined, they suggest that rhetoricians analyze that knowledge through language: "By working with what we *can* figure
out—that is, by emphasizing the rhetoricity of these scientific arguments *as
we import them*—we combat their tendency to overwhelm and, by troubling
default, resume their status as foundational knowledge."[19] Jenell Johnson and

---

15. Gorgias, *Encomium of Helen,* 53.

16. Gorgias, 53.

17. Rickert, *Ambient Rhetoric.*

18. Mays and Jung, "Priming Terministic Inquiry," 42, 47.

19. Mays and Jung, 47, emphasis in original.

Melissa Littlefield raise similar concerns about the adoption of neuroscience findings without attention to how those findings are presented rhetorically in popularizations.[20] Like these scholars, I share an interest in the rhetoricity of neuroscience texts. The goal of my inquiry, however, is to allow for a deeper engagement between the neurosciences and the humanities. Focusing only on rhetoric-as-persuasion, text, or argument still blackboxes much of the process through which scientific knowledge claims are generated in neuroscience. This book seeks to pry open that black box to show the rhetorical-material considerations that go into neuroscience imaging experiments themselves. In so doing, I identify ways that humanistic insights can problematize, enrich, or extend neuroscience findings.

A second strand of neurorhetoric draws on neuroscience research to theorize rhetoric. Researchers in this vein often engage theories of embodied cognition, mirror neurons, and affect, all of which allow them to argue for ways that rhetoric might work outside of the symbolic, how it might act on and through bodies.[21] While I am sympathetic to these moves, I would argue that it is important to pay attention to how neuroscientists enact such concepts and to whether they square with our own understandings. I agree with Victoria Pitts-Taylor, who argues that "materialism does not absolve the need to critically assess neuroscientific knowledge and practice" or the need to "question epistemic claims" rather than seeing neuroscience as a "neutral lens through which to see the brain and nervous systems."[22] As Pitts-Taylor puts it, we must not lose sight of how scientific practices "enact their objects of investigation and have material effects on the bodies they study."[23] Rhetoric scholars, especially, must not ignore their own scholarly tools when turning to neuroscience but should bring those tools along to engage critically with that material.

By critically attending to the site of the experiment, I extend research in rhetoric dating back to the 1980s. In 1988, for instance, Mike Rose drew attention to the "cognitive reductionism" that often shapes descriptions of the challenges that basic writers face with cognitive or neurological deficiencies. This tendency, he argues, can "lead to social distinctions that have important consequences, political as well as educational."[24] Notably, Rose does not seek to dismiss cognitive insights out of hand, but simply argues that the rich potential of those insights "should not keep us from careful consideration

20. Johnson and Littlefield, "Lost and Found in Translation."

21. See Gruber, "Reinventing the Brain, Revising Neurorhetorics," and Gibbons, "A Neurorhetoric of Incongruity."

22. Pitts-Taylor, The Brain's Body, 11.

23. Pitts-Taylor, 11–2.

24. Rose, "Narrowing the Mind and Page," 268.

of their limits, internal contradictions, and attendant critical discussion and counterstatements."[25] He notes, for instance, the considerable technical challenges posed by electroencephalogram (EEG) techniques, then ascendant in the field of neuroscience, which included questions about electrode placement, data analysis, and the kinds of stimuli used.[26] As we will see, similar problems abound in fMRI-based research.

While Rose employs these methods primarily to question problematic uses of cognitive science, Walker models a more affirmative (but still careful) approach in "Of Brains and Rhetorics," published in 1990. While the scientific debate over brain lateralization that Walker discusses has long since been put to rest, Walker's approach calls for "a healthy dose of skepticism" and an "equally healthy exercise of the rhetoric of assent," or, he says, "at least a survey of the grounds of possible assent."[27] Walker uses this exercise to argue that neuroscience evidence seemed to be converging around an Aristotelian understanding of cognition. Importantly, though, he also argues that researchers in neuroscience might do well to return to Aristotle's thought for a richer conception of cognition in which "reasoning, imagination, feeling, and language (in its various potentialities) are inextricably joined in normal cognition."[28] Indeed, at least some research in social and cognitive neuroscience is now embracing this viewpoint, as I will show in chapter 6.

Daniel Gross's book, *The Secret History of Emotion*, follows from Walker's argument that Aristotelian theories of mind and cognition prove fruitful for modern brain science as well as modern rhetoric. Although he begins with Aristotle, Gross identifies concepts that stress the social dimensions of emotion across historical periods. Gross argues that scholars in the human sciences might learn from the rhetorical tradition "for insight into the emotions," especially given that rhetoric "was the first, and remains the richest, resource for such inquiry."[29] Gross's book illustrates the value of humanities scholarship to neurosciences even as scholars in the humanities look to neuroscience for their own inventional work. It is in that spirit, raveling back and forward between the two, that this book will proceed. In the next section, I will out-

---

25. Rose, "Narrowing the Mind and Page," 268.

26. Rose, "Narrowing the Mind and Page," 278. On the limits of brain lateralization theories from a composition perspective, see also Brueggemann, "Whole Brains, Half-Brains, and Writing."

27. Walker, 303–4.

28. Importantly, however, this model of cognition continues to fluctuate with what Walker calls the Neoclassical/Romantic view in which logic and emotion are divided, particularly in fMRI-type studies that require neuroscientists to study a single purified psychological concept at a time. Walker, 314.

29. Gross, *The Secret History of Emotion*, xvii.

line how I draw from rhetorical studies of science and science studies research (and especially feminist scholarship) to ground my approach.

METHODOLOGY

This project employs insights drawn from the rhetoric of science and technology. Namely, I assume that material, linguistic, and social aspects of science are intertwined, or *raveled*. The title of this book, *Raveling the Brain*, evokes that intertwinement, and I use the term *ravel* in three ways throughout the book. First, I use the idea of *raveling out* as a way of puzzling out a problem. This suggests that the brain is not a single, universal entity that can be definitively pinned down by any one research method. Instead, I take the view that ontologies of the brain are multiple, enacted differently in different times and places and through different research methods, so they must be puzzled or "raveled out" through multiple perspectives. Second, I follow a process of *raveling back*, pulling on the strands that currently inform our understanding of the brain and working back to see how they were previously knotted and entangled. Using rhetorical genealogy, I work backward to see how previous understandings of the brain have emerged and receded as new tools, methods, concepts, and arguments are progressively woven over time. Third, I see myself *raveling together* humanistic theories (and especially rhetorical theory) with the disciplines that make up the neurosciences. Rhetorical theory, in particular, is an untapped resource that can enrich scientific studies of the brain by "multiplying" the range of considerations that neuroscience research studies. In what follows I will briefly describe each of these senses of "raveling" and how they will be used in ensuing chapters.

Raveling Out

Scholars in rhetoric now echo the perspective taken in science studies more broadly—often inspired by feminist research—that material, social, and linguistic elements interact, inseparably, in the study of scientific objects. As Celeste Condit puts it, "the truth of a scientific claim is a product not of the interface of an objective language with an objective material reality—nor merely a product of discursively mediated social forces—but rather it is the interface of a socially imbricated linguistic code structure with a selected material reality."[30] Condit, however, suggests that the material and linguistic

---

30. Condit, "How Bad Science Stays That Way," 86.

elements of scientific practice can be separated when she writes that "understanding the social biases we harbor can help us to recognize linguistic biases in research and produce scientific findings less hobbled by linguistic shortcomings, and thus more appropriate for our frames and interests."[31] Others would argue that disentangling the two is not so easy. Karen Barad, for instance, puts it this way: "Measurement is a meeting of the 'natural' and the 'social.' It is a potent moment in the construction of scientific knowledge—it is an instance where matter and meaning meet in a very literal sense."[32] While Condit tends to emphasize rhetoricians' role in terms of the linguistic, Barad would insist that the linguistic and material are not easy to disentangle. Like Condit, then, I advocate for a "broad program" of rhetoric of science, one that "requires that one understand sufficiently the scientific claims being made qua science," yet I do not believe the interest of rhetoricians should rest mainly on the level of "linguistic bias" or "linguistic short-comings" but should also consider the material-semiotic enactments of concepts that, in the case of neurosciences, can also be understood from humanistic perspectives.[33]

Thus, I hold, along with Barad and others, that the material and linguistic cannot be separated; they are intertwined. For example, the language that brain scientists use to label the concepts they are interested in measuring (such as pleasure or empathy) is also embedded in the tools they use (from psychometric tests to Skinner boxes to stimuli used in an fMRI experiment) and, therefore, to how those concepts materialize in the brain. Rhetorical analysis allows us to "ravel" those threads in the sense of explicating or puzzling out how they are connected. The tools of rhetorical analysis are well suited precisely for that purpose because they allow us to identify particular configurations of material-discursive elements. I will show how rhetorical tropes (such as metaphor and synecdoche) can be understood not just as linguistic features of a text but as tools for understanding how psychological concepts are enacted in scientific experiments. These tropes form part of what I am calling the "rhetorical-material infrastructure" of neuroscience, or the way that rhetorical figures, arguments, and topoi become entwined with scientific practices and tools, each enabling the other. Identifying them helps us to tease out how the various threads making up contemporary neuroscience are woven together.

---

31. Condit, "How Bad Science Stays That Way," 87.
32. Barad, *Meeting the Universe Halfway,* 67.
33. Condit, "How Bad Science Stays That Way," 87.

Raveling Back

Feminist science studies scholars have questioned the notion that scientists make disinterested, objective observations of the world. This ideal, as Barad argues, reflects classical assumptions about measurement that do not stand up to the more recent findings of quantum physics, which suggest that "*the nature of the observed phenomenon changes with corresponding changes in the apparatus.*"[34] The classical view assumes that individual objects and abstract concepts can be clearly isolated and that measurement of those objects can be separated from observation.

Particularly significant for Barad's theory of agential realism are apparatuses. Unlike a classical view that positions apparatuses as necessary to, but not constitutive of, scientific observations, in Barad's view apparatuses participate in enacting scientific observations. According to Barad, we cannot think of scientific observations as simply involving humans who measure a concrete reality without changing what is observed; agential realism suggests, instead, that scientific observations emerge from specific material-discursive arrangements that include measurement itself.[35] For Barad, apparatuses are "macroscopic material arrangements through which particular concepts are given definition."[36] This is very clearly the case in social and cognitive neuroscience experiments that use fMRI, because researchers must make decisions about how, exactly, to define and then operationalize whatever it is they plan to study. As I will show in the following chapters, those decisions become part of the apparatus used to study the phenomenon in question; in fact, the apparatus constitutes or enacts that phenomenon. Thus, we must include in our consideration of phenomena (as they are enacted via apparatuses) "all relevant features of the experimental arrangement."[37] Much like rhetoric involves, in Aristotle's terms, "the faculty of observing, in any given case, the available means of persuasion,"[38] a Baradian approach to analyzing scientific experiments would involve the "faculty of observing, in any given experiment, the relevant features of the experimental arrangement."

In the case of social and cognitive neuroscience, we must cast a wide net in considering these "relevant features." A social and cognitive neuroscience experiment does not measure a concrete "thing" (pleasure, or creativity, or politics) but enacts a phenomenon through particular practices and configu-

---

34. Barad, 106, emphasis in original.
35. Barad, 142.
36. Barad, 142.
37. Barad, 119.
38. Aristotle, *Rhetoric*, 24.

rations that actively *materialize* that which they purport to study. An fMRI machine is part of an apparatus for materializing the brain, in other words, but that apparatus is not confined to the fMRI machine itself, or even to the space-time of the laboratory, but extends into the world.

Barad shows that relevant features of a phenomenon may include the practices of gender, race, and class; she gives the example of an experiment that Otto Stern and Walther Gerlach employed in the 1920s in Germany—now considered canonical—that includes a bad cigar that left Stern's breath laced with sulfur which, in turn, made the traces of silver sulfide on plates more readily visible. The cigar, Barad argues, is a "condensation" of "the workings of other apparatuses, including class, nationalism, economics, and gender," and all of these, she claims, should be considered part of the scientific apparatus for that experiment.[39] This is also the case for social and cognitive neuroscience, as I will argue here; in fact, these entanglements are rather obvious once one considers that the "apparatus" for an fMRI experiment includes the scientists involved, participants chosen, the means used to select them, the stimuli used and how they are measured, the statistical methods used to analyze results (including various software packages and tools), and so on. Barad argues, in particular, that concepts are not simply "ideational in character; they *are specific physical arrangements.*"[40] This point is especially important for a discipline that purports to connect psychological concepts with their neural correlates. As I show in chapter 2, to do so, neuroscientists currently take those concepts to be real things that exist prior to the experiment rather than considering them as "*specific physical arrangements.*" Yet, in an fMRI experiment, a concept like attention or creativity or pleasure is enacted through material arrangements of the various elements I listed above: psychological concepts materialize through the arrangements of these features. I will argue in chapter 2 that they often materialize in ways that are necessarily reductive because of the limitations of fMRI machines and the standard experimental procedures used in the discipline.

To determine how those arrangements come together, I employ the strategy of "raveling back," employing rhetorical genealogy to move back in time to see how current material-discursive practices are woven out of earlier ones. I draw the idea of "rhetorical genealogy" from Mary Queen, who describes it as a process of analyzing "the cultural practices and rhetorics through which particular representations and interpretations gain validity and power."[41] In the case of social and cognitive neuroscience, these cultural practices include

39. Barad, 164–65.
40. Barad, 109.
41. Queen, "Transnational Feminist Rhetorics," 476.

dominant scientific paradigms for understanding human minds and culture that have helped the discipline to gain scientific authority.

However, I am extending Queen's term to include the materials through which neuroscience is enacted. Thus, I am calling my method "rhetorical-material genealogy," which includes attention to the tools, practices, and techniques of neuroscience that are intertwined with the fields' discursive features. While it is important to analyze neuroscience claims and texts rhetorically, I will demonstrate throughout this book that our analysis need not rest at the level of textual arguments. Drawing on the methods of Bruno Latour and Barad, I will show how we can work through scientific language—especially the language used in methods sections of scientific articles—to reconstruct how scientific experiments themselves operationalize behavioral and psychological concepts in order to render them measurable in fMRI settings. In making choices about what to study, with which participants, which stimuli, and which hypotheses and assumptions, I will argue, neuroscientists necessarily draw on a set of sedimented practices, tools, and concepts that allow them to make knowledge claims in the first place. This rhetorical-material infrastructure works largely through neurorealism (assuming that concepts such as pleasure or creativity are concrete, definable things) and neuroessentialism (assuming that those concepts can be reduced to a quantifiable entity that can be operationalized within an fMRI setting). However, it also draws on culturally available traditions and practices informing how those concepts should be understood. By combining this method with rhetorical genealogy, I show how neuroscientists are (consciously or not) drawing on embedded traditions that enable these ways of thinking about concepts such as pleasure, creativity, violence, politics, gender, sex, and persuasion.

Raveling Together

Identifying these features of fMRI neuroscience research is not meant simply as a critique or debunking move but as a lever we can use to show how humanistic research can work together with neuroscientific concepts. By intertwining, or "raveling together," neuroscience and rhetoric, both disciplines stand to benefit. Consider how the strength of a rope depends on the number of strands or threads that are entwined together: the more strands, the stronger the rope. In the same way, raveling neuroscience and rhetoric together leads to a stronger approach to understanding brains. Humanistic theories, I argue, can help to 1) provide a richer account of the relevant features of a phenomenon by drawing attention to features that go unnoticed in scientific accounts and 2) suggest alternative configurations for fMRI experiments that would pay closer attention to relevant features that do not always undergo scrutiny.

In this analysis, I am less interested in rhetoric as a way of identifying "errors" or sources of bias than in thinking through the knowledge practices of neuroscientists and how they differ from (or could benefit from) those of humanities scholars. Since we each have our "terministic screens," we each stand to benefit from interactions. Kenneth Burke famously calls attention to the way that the terms we use necessarily "direct the attention to one field rather than to another."[42] In a sense, Burke went on to say, the observations we make (either in our daily lives, or in this case as researchers in a particular discipline) are *"but implications of the particular terminology in terms of which observations are made."*[43] I will show how, for instance, viewing the act of composition through the lens of "creativity" (in neuroscience) versus "invention" (in rhetoric) leads to different kinds of observations—as well as different kinds of questions. By intertwining these two perspectives, neuroscientists and rhetoricians might open up new lines of inquiry or new ways of theorizing human cognition.

Yet, a power differential between the two fields (in terms of prestige, funding, etc.) means that the humanities are often positioned as less influential. As Barad puts it, "What is needed are respectful engagements with different disciplinary practices, not coarse-grained portrayals that make caricatures of another discipline from some position outside it."[44] In other words, the goal is not to privilege one or the other but to put them into critical conversation with each other. My hope is that the analysis that follows will be interpreted as "respectful engagement" in the cognitive and social neurosciences from the perspective of a humanities scholar.

In what follows, I employ rhetorical analysis (not only of articles but of experiments themselves) to examine concepts studied neuroscientifically that can also be understood through rhetorical theory. I use rhetoric scholarship (and other humanistic scholarship) to offer ways of addressing the problems posed by neurorealism and neuroessentialism. Drawing on traditions of humanistic thought, especially rhetorical thought, I offer alternative ways of understanding the concepts in question that blend neuroscientific theories with rhetorical ones.

Finally, I conclude each chapter with a move that David Gruber calls "mobilization": that is, suggesting paths forward for rhetorical engagement

---

42. Burke, *Language*, 46.
43. Burke, *Language*, 46, emphasis in original.
44. Barad, *Language*, 93.

with these concepts and with rhetorical cognition more broadly.[45] I argue that the rhetorical-material infrastructure of the neurosciences could be enriched if it drew more deeply on humanistic theories and methods, particularly rhetoric. While neuroscience advertises itself as an interdisciplinary field, drawing on biology, physics, engineering, and psychology (to name a few), to date it has engaged in more limited ways with the humanities. A humanistic perspective (and especially a rhetorical perspective) illustrates the material-discursive infrastructure shaping neuroscience research. It also generates alternative models that can address some of the limitations of current neuroscientific practice. Yet, the argument can also be turned on its head—with caution—to suggest how the humanities (and in my case, rhetoric), can learn from the neurosciences, and I point out implications that work in that direction in each chapter.

OUTLINE OF CHAPTERS

Chapter 1 examines how it became possible to locate complex emotional and psychological phenomenon in specific brain regions. Taking up neurorealism claims, I show how arguments that localize mental concepts in the brain occur mainly at the level of the first stasis of rhetoric, conjecture (which answers the question "Does it exist?"). To determine which brain region corresponds with, say, empathy or reason, researchers first have to reify the concept in question, arguing it into place as a concrete thing. To do so, they must also reify the brain using metaphors such as that of "brain as computer" or "brain as a network." Neurorealism, then, is not simply a product of popularization but an important point of invention, or heuristic, that is required for the knowledge contemporary neuroscientists seek to generate.

Chapter 2 examines how it became possible to reduce mental phenomena to the localized brain responses measured by electrical stimulation, neuroimaging technologies, and other techniques. Taking up neuroessentialism, this chapter investigates how arguments equating mental concepts with brain activity or regions operate at the stasis of definition. In order to define mental concepts in this way, neuroscientists must reduce them so that they can be operationalized within particular research methods. Here, I use the example of "creativity," showing how neuroscientists must constrain that term to make

---

45. Gruber, "Reinventing the Brain, Revising Neurorhetorics," 247.

it measurable. I show how a rhetorical understanding of "creativity" as "invention" offers a richer viewpoint from which to study this phenomenon.

Chapter 3, "Neurorhetorics," examines how neuroscientists and rhetoricians study the phenomenon of persuasion. In fMRI studies, neuroscientists typically treat persuasion as a matter of discrete, rational events in which individuals are "exposed" to an argument and then are either convinced or not. Similar assumptions guide scientific studies of the persuasiveness of neuroscience itself, of which there have been quite a few. Instead, I suggest, rhetorical theories would push neuroscientists to understand persuasion as occurring over time, through repeated, serialized encounters with arguments, tropes, topoi, and affects.

The next three chapters demonstrate what a neurorhetorical approach might look like, showing how humanistic concepts and theories can supplement neuroscientific accounts of sex/gender, politics, and persuasion. In chapter 4, "Neurosex," I take up research about sex and the brain, particularly in relation to gender and sexuality. Focusing on the rhetorical-material infrastructure of neuroscience helps to explain why research advocating seemingly retrograde assumptions about men and women, homosexuals and heterosexuals continue to be published, but it also allows me to identify places where feminist insights have been integrated into neuroscience research. That research, while by no means common, nonetheless offers possibilities for studying sexed and gendered brains differently, in ways that call into question views of hardwired, innate sex differences.

Chapter 5 engages "Neuropolitics," or claims about the neurological basis of political orientations. Here, I examine how these studies employ rhetorical commonplaces of "right" and "left," or conservative and liberal, politics, commonplaces that themselves shape what we understand politics to be and hence the kinds of neurological findings that are unveiled. In contrast to these binaristic ways of understanding political ideology, I suggest that rhetorical theories of politics offer a more fluid understanding how political ideologies arise out of circulating, material-semiotic rhetorics and affects. Current research in neuroscience offers some glimmers of possibility, questioning whether political behavior results not simply from innate brain structures or genetics but possibly reflects political action within a specific type of political context.

Chapter 6 examines "Neuroaffect," considering how researchers in neuroscience and rhetoric have both relied upon a classical psychology view of affect in their attempts to understand how feelings circulate through the body, brain, and society. A strand of research in the neuroscience of emotion, however, seeks to consider emotions as linguistic constructs, and this type of research offers a model for a rich study of affect that would ravel together humanistic and scientific insights.

The concluding chapter argues that neuroclaims arise from neuroscience itself because they serve inventive purposes for researchers. These stases inform not only how neuroscience is sold to public audiences but how it is conducted at the level of research methods, theories, and practices. While examples of misappropriation abound, in some cases researchers do model how interdisciplinary perspectives can enhance neuroscience, leading toward richer, more productive approaches to building knowledge about the brain and its role in human culture. The book concludes by summarizing some of the things that we can currently say about neuroscience and rhetoric and by raising questions that remain unanswered, as well as identifying future research directions for investigations of rhetorical cognition.

CHAPTER 1

Neurorealism

THE BRAIN is big business. While neuroscience researchers profess uncertainty about the particulars of how our brains work, this has not stopped entrepreneurs from seeking to profit from the technologies neuroscientists use—such as fMRI, EEG, eye-tracking, and biometric responses—to understand what makes consumers tick. To do so, they posit that there must be some particular brain region that can reliably indicate one's response to a product or an advertisement, or what some neuromarketing experts refer to as the brain's "buy button." For example, a report in an online magazine described how Frito-Lay repackaged one snack, Sun Chips, to appeal to women by not tripping "the brain's 'junk food' switch."[1] Yet it is not only (questionably scrupulous) neuromarketing wannabes who go around talking like this. Indeed, the term *buy button* has appeared in the titles of scientific articles, suggesting that neuroscience researchers, too, share the belief that specific brain regions correlate with shopping behavior.[2] (Take, for instance, an article in *NeuroImage* entitled "Multiple 'Buy Buttons' in the Brain.") At the same time, both popular news articles and scientific articles—*the same ones* that refer to the "buy button"—will simultaneously insist that there is no buy button in the brain, that the brain is so complex that choosing one product over another

---

1. Zara Stone, "The Scientists Who Control the Brain's 'Buy' Button," OZY.com.
2. Simone Kühn, Enrique Strelow, and Jurgen Gallinat, "Multiple 'Buy Buttons' in the Brain."

likely involves multiple processes or systems. In either case, the idea of a "buy button"—its existence or lack thereof—depends on the assumption that neuroscience research can uncover some kind of trace or mark of a behavioral phenomenon that exists in the brain. Why is this the case?

One answer might be simply to note that science works by reducing the materiality of experimental practice to a representative trace. This is the key point that Bruno Latour and Steve Woolgar make in *Laboratory Life,* their ethnographic account of scientists studying rat brains and how those brains (along with the various people, tools, and processes used to study them) are condensed into a visual trace, a graph, that is featured in a scientific article while the rest of those elements are elided. The trace is substituted for that broader assemblage of elements. Here, though, I wish to offer a rhetorical genealogy of how this kind of substitution takes place in neuroscience and how it is related to neurorealism. According to Eric Racine, Ofek Bar-Ilan, and Judy Illes, "neuro-realism" occurs when "coverage of fMRI investigations . . . make[s] a phenomenon uncritically real, objective or effective in the eyes of the public."[3] Yet, I would argue that the problem goes deeper than popular news coverage, to the very roots of neuroscience investigation into human mental and emotional concepts, to the intra-actions of apparatus, theory, and language that ground the neurosciences. In particular, neuroscience research relies on several guiding metaphors that function as terministic screens, reflecting and deflecting researchers' attention toward and away from certain aspects of cognition.

In this chapter, I first demonstrate how neurorealism works rhetorically by drawing on the rhetorical stasis of fact and the rhetorical figure of metaphor. Next, I provide a genealogy of some of the key metaphors that have been used to understand how qualities and practices can exist in the brain; these metaphors are not striated into distinct historical periods but instead percolate through time so that, in the present, we have resonances of multiple metaphors that seek to fix ephemeral qualities within the brain.[4] Finally, I show how this history makes apparent currently dominant metaphors that figure the brain as a computer, a figure especially central to fMRI research in social and cognitive neuroscience. In the case of decision-making, this metaphor coincides with ableist and capitalist ways of understanding the mind that suggest that the brain is a high-functioning machine for calculating risk and benefit. A rhetorical perspective challenges neuroscientists to think through the implications of this metaphor. Ultimately, this chapter shows the centrality

---

3. Racine, Bar-Ilan, and Illes, "fMRI in the Public Eye," 160.
4. Jensen, "An Ecological Turn in Rhetoric of Health Scholarship," 524.

of metaphor to neurorealism and the centrality of neurorealism to contemporary social and cognitive neuroscience that uses fMRI, and it shows how investigating neurorealism by focusing on metaphors can serve as a way of reading neuroscience and inquiring into its rhetorical workings.

HOW NEUROREALISM WORKS

Neurorealism claims operate at the stasis of fact, answering the question "What is it? Does it exist?" A claim based on neurorealism asserts that a mental concept or quality (such as memory, love, or jealousy) is, to borrow the terms used by Racine, Bar-Ilan, and Illes, "uncritically real," often by arguing for a specific location for that concept in the brain. The rhetorical strategy that produces reification is metaphor, a term derived from the Greek metaphorá, meaning "transfer." A metaphor might be understood as a trope that transfers meaning from one term to another; more specifically, according to Kenneth Burke, the basic strategy involved in metaphor is "to convey some incorporeal state in terms of the corporeal or tangible."[5] For example, if someone says "my feet were a block of ice," he or she is trying to convey the sensation of cold feet (an intangible sensation) in terms of something concrete, an actual block of ice. This is one way that neurorealistic metaphors work: they serve to concretize the brain, something that is still far too complex to be rendered tangible, in terms more readily understood.

Yet, metaphors also relate to terministic screens in the sense that they focus attention on some aspects of a phenomenon and not others. They are framing devices. As Burke puts it: "Metaphor is a device for seeing something *in terms of* something else. It brings out the thisness of that, or the thatness of this."[6] Ken Baake uses music as a metaphor to explain this, arguing that metaphors themselves are akin to musical notes, emitting "sounds" that reverberate outward, carrying different implications or signals along with them. Some of these signals may be more appropriate to a given scientific discipline or discourse community than others.[7] What is transferred in a metaphor is not merely a name but the predicates or associations that accompany it. For instance, understanding the brain in terms of a map leads researchers to plot out key "regions" or "locations" as well as "seats," or centers for certain capacities. Research often proceeds by carrying out the implications of a central metaphor.

---

5. Burke, "Four Master Tropes," 424.
6. Burke, "Four Master Tropes," 421–22.
7. Baake, *Metaphor,* 7.

The key functions of neurorealist metaphors are substitution and reification. Neurorealistic metaphors convey material states or behaviors (the dispositions and actions of various people, things, and ideas that are enacted in an experiment) in terms of neurological and neurochemical processes (substituting one for the other) and then reify the substitutions as real. For instance, say a study of decision-making (a behavior) links that process to a certain brain region (the prefrontal cortex and the amygdala). That brain region then comes to substitute for the process itself and might be posited as a "decision-making center." That center is then posited as a real thing, an actual region in the brain that *really does* process decision-making. The "decision-making center" is a metaphor, but it is one that comes to take on the guise of a real, concrete thing. The fundamental function of metaphor—to transfer or carry—can move meaning not just from the incorporeal to the corporeal, but from the corporeal to the neurochemical. The transfer substitutes a broader set of things for something else that is considered "more real."

I'm not the first to note the centrality of metaphors to neuroscience. Metaphors can reify any number of psychological constructs. According to David E. Leary, a psychologist, "Qualities, events, and any other aspect of experience are included among the innumerable 'things' that can be rendered through metaphor."[8] Then, neuroscientists track down the implications of seeing the brain *in terms of* the metaphors they use: in terms of a computer, or a network, or of a system of governance. Leary shows how the terms from one metaphor extend into research practices and assumptions. For instance, Leary argues that "when Aristotle treated the mind as a living thing, he invited the inference that it can develop and change over time, and when cyberneticists make information central to biological functioning, they set the stage for questions about the relationship between the 'noise' and 'messages' involved in the regulation for living bodies."[9] Each metaphor guided scientists toward certain kinds of investigations and away from others.

Such uses of metaphor do not merely explain or illuminate things. Instead, metaphor serves inventive purposes. According to Burke, "By deliberate coaching and criticism of the perspective process, characters can be considered tentatively, in terms of other characters, for experimental or heuristic purposes."[10] That is, by considering one thing in terms of something else, one can generate new ideas, lines of argument, theories, or concepts. Jeanne Fahnestock notes that "a metaphor need have no previous or easily categorized link to the words it replaces or joins. Instead it creates new links, allowing the

---

8. Leary, *Metaphors in the History of Psychology,* 5.
9. Leary, 5.
10. Burke, "Four Master Tropes," 422.

rhetor to illuminate one term (or concept) by features or senses borrowed from another."[11] In the case of neuroscience, the metaphors used to understand the brain have helped to create new theories, concepts, and lines of research. While other scholars, such as David Gruber, position neurorealism primarily as a rhetorical strategy,[12] I argue that it also works much more deeply as a grounding principle of the discipline. It is not only a textual feature but also represents a disciplinary orientation of the brain that materializes in research methods, tools, and apparatus. Investigating a discipline's guiding metaphors offers a way of identifying its workings. That is, guiding metaphors are instantiated in research practices, tools, and apparatus, as well as in its language, all of which intra-act in any scientific study.

The next section provides a short history of some of the metaphors that have informed neuroscience research over time. This history is not meant to be exhaustive but instead to suggest how, at different times and places, metaphors have informed how we understand the brain and how neuroscience research has been materially practiced. My point is that neurorealism advanced via a series of metaphors that seemed to make localization more scientifically sound while assuming a one-to-one correlation between a mental state and a brain region. Neurorealism has a long history as endemic to the scientific study of the brain; it's not just a feature of popularization. And while neurorealism certainly has its limitations, it has proved essential for the study of the brain to progress at all.

A RHETORICAL HISTORY OF NEUROREALISM

The idea that the mind or soul has a concrete location in the body harks back to antiquity. In ancient Greece, philosophers and natural scientists questioned the location of the ruling principle (*hegemonikon*) of the body. Some, like Aristotle, thought it must be in the heart; others, like Plato, thought it lay in the head; still others positioned it in the diaphragm or lungs, since breath and air were presumed to constitute the human life force or soul.[13] The term *hegemonikon* draws on the metaphor of the ruler, commander, or leader: just

---

11. Fahnestock, *Rhetorical Style*, 105.

12. Gruber, "Three Forms of Neuro-Realism," 194.

13. Rocca, *Galen on the Brain*, 19. Richard Onians notes that in Homer, "thinking is described as 'speaking' and is located sometimes in the heart" but often in an area "traditionally interpreted as the 'midriff' or 'diaphragm'" (*Origins of European Thought*, 13) but possibly more aptly translated as "lungs" (29).

as one city-state rules over those surrounding it, the *hegemonikon* (whether positioned in the lungs, head, or heart) governs the body.

Among the Ancient Greeks, others clearly located the functions of the soul within the brain. In "The Sacred Disease," a text attributed to Hippocrates (460–c. 370 BCE), we see an insistence that "from the brain, and from the brain only, arise our pleasures, joys, laughter, and jests, as well as our sorrows, pains, griefs and tears."[14] The Greek physiologist Herophilus (335–280 BCE) also located the soul in the brain, locating different types of thought in the ventricles of the brain, again using a tripartite scheme: he located judgment in the anterior (or lateral) ventricle, cognition in the middle (or third) ventricle, and memory in the posterior (or fourth) ventricle.[15] In Galen (129–210 ACE), too, we see a tripartite set of key mental functions located in the brain: imagination, reason, and memory.[16] Later, in the fourth century AD, Poseidonius sought to localize those mental functions in the front, middle, and back of the brain, respectively.[17]

In ancient Greece, then, the soul was not yet located uniformly in the head or brain. It was, however, common to divide the soul into its constituent parts by extending the logic of the *hegemonikon* and drawing on the rhetorical trope of *divisio*. Rather than one single ruler, philosophers and natural scientists considered whether the body might be governed by a series of rulers, each responsible for different functions (much as a city-state may be divided into a series of classes or groups). Consider how Plato's *Timaeus* outlines a tripartite theory of the soul, wherein each part corresponded to one of the parts of Plato's ideal society. The first part of the soul, responsible for wisdom and reason, was lodged in the head, and corresponded to the guardian class of society. The second, responsible for passion and courage, was housed in the thorax, and corresponded to the warrior class. And the third part of the soul, responsible for avarice and cupidity, was located in the abdomen, and corresponded to the proletariat class.[18] Indeed, in the *Republic*, Socrates states that "there are the same classes, and the same number of each, in the state as in the soul of each individual."[19] Here we have a clear example of mental concepts ascribed to the soul or mind tied explicitly to social and political status. The metaphor of the *hegemonikon* and the city-state guides Plato's understanding of the soul.

---

14. Hippocrates, "The Sacred Disease," 175.
15. Rose, "Cerebral Localization in Antiquity," 243.
16. Rose, "Cerebral Localization in Antiquity," 244.
17. Rose, "Cerebral Localization in Antiquity," 245.
18. Smith, "The Triune Brain in Antiquity," 3.
19. Plato, *Republic*, 427.

Plato's understanding is similar to that taken up much later by phrenologists, who amplified the number of divisions greatly but took the same idea of partition, with regions of the brain (or organs) being responsible for different functions. We might be tempted to construct a progress narrative from a Platonic understanding forward in time, but this compartmentalized understanding of the brain has not always held sway. For instance, Aristotle viewed the soul as infused throughout the body, but with the heart as its center. Like Plato, he viewed the soul as tripartite, consisting of the vegetative, sensitive, and ratiocinative soul.[20] Yet, the connection between the soul and the body could be seen, Aristotle held, in how the emotions were embodied: "It therefore seems that all the affections of soul involve a body-passion, gentleness, fear, pity, courage, joy, loving, and hating; in all these there is a concurrent affection of the body. . . . Consequently their definitions ought to correspond, e.g. anger should be defined as a certain mode of movement of such and such a body (or part or faculty of a body) by this or that cause and for this or that end."[21] Here is a conception of emotion rooted not in the head or brain but diffused throughout the body and its movements, one that might anticipate the embodied cognition theories that attract humanities and social theorists (such as Andy Clark's "extended mind" hypothesis).

In all of these ancient thinkers, we find an early version of realism, if not neurorealism specifically: that is, regardless of whether they located mental concepts throughout the body, in the heart, the brain, or in specific brain locations, these thinkers took for granted that those mental concepts were real, concrete things locatable in the body. In the Enlightenment period, philosophers and natural scientists continued to inquire into the location of the soul or mind, extending the Galenic theories described above and grounding them in incipient knowledge gained from examining preserved brains. John Locke, for instance, insisted that the brain was the "seat of sensation" that received "animal spirits" via the nerves.[22] Locke gives the following example: "Fire may burn our bodies, with no other effect than it does a billet, unless the motion be continued to the brain; and there the sense of heat, or *idea* of pain, be produced in the mind wherein consists *actual perception*."[23] Locke relied in part on the dissections of the human brain completed by Thomas Willis, whose *Cerebri Anatome* (1664) illustrated new methods of preserving brains and dyeing them.[24]

---

20. Smith, "The Triune Brain in Antiquity," 9.
21. Aristotle and Roberts, *Rhetoric*, 1.1.
22. Locke, *An Essay Concerning Human Understanding*, 2.8.12.
23. Locke, 2.9.3.
24. Willis, *Cerebri Anatome*.

For our purposes, though, it is instructive to turn to the nineteenth century, which witnessed a different way of understanding the brain that occurred amid scientific developments (such as the rise of cranial comparative anatomy and renewed interest in physiology) and concurrent social changes (including the debate over slavery and the so-called "woman question").[25] Instead of a tripartite division of the mind, in the nineteenth century it became commonplace to understand the brain as a larger set of organs. Here, the driving metaphor is that the brain is a body, with many different organs, each with a specialized function. Phrenologists argued that these organs could be identified externally by using a craniometer to map the various bumps and protrusions on a person's skull, with the idea being that internal brain organs left their mark externally, on the outside of the head. The scheme of *divisio* appears again here, but so does amplification: instead of three parts, phrenologists amplified the list to include upwards of thirty separate functions, from "amativeness" to "carnivorousness."

Today, the term *phrenology* is synonymous with quackery, junk science, pseudoscience, and charlatanism. If it is evoked at all in scientific discourse, it is as an epithet or warning. In an article published in *Science*, Robert T. Knight lauds a series of new studies because they "dispel any phrenological notion that a given innate mental faculty is based solely in just one part of the brain."[26] Phrenology has come to represent what current neuroscience is not, or should not be. However, the influence of phrenology on the rhetoric of psychology and what is now called neuroscience may be greater than some would like to admit. Neurosurgeon Donald Simpson writes that even after phrenology was debunked as fraudulent, "the phrenological idea of cerebral localization remained, and was a force in the great neurological discoveries that were made later in the nineteenth century."[27] Localization literalizes or reifies mental concepts in the brain. Phrenology laid out *rhetorical* groundwork for reification that still holds today.

Though often termed the father of phrenology, Joseph Franz Gall did not actually like that term, preferring to call his doctrine "functions of the brain" or "brain physiology," or, later, "organology."[28] The latter term most clearly references the central metaphor informing his theory: the idea that the brain was like a body, with distinct organs, each performing a discrete function. In *Sur l'origine des qualité morales et des facultés intellectuelles* (1822), he enumerated four principles of his doctrine:

---

25. Shapin, "The Politics of Observation," 142–43.
26. Knight, "Neural Networks Debunk Phrenology," 1578.
27. Simpson, "Phrenology and the Neurosciences," 475.
28. Clarke and Jacyna, *Nineteenth-Century Origins of Neuroscientific Concepts*, 222.

1. "That the moral qualities and intellectual faculties are innate"
2. "That their exercise or their manifestation depends on the organization [of the brain]"
3. "That the brain is the organ of all penchants, sentiments, and faculties"
4. "That the brain is composed of as many specific organs as there are different tendencies or faculties."[29]

Gall's rhetorical innovation did not lie in the use of the metaphor of the organ, which had already been used by Swiss biologist Charles Bonnet in 1770 (whom Gall had read and cited) and by Pierre Jean George Cabanis.[30] Instead, Gall's innovation was in carrying out the implications of that metaphor to a new degree: namely, he elaborated the multiple "organs" that must exist if we understand the brain not as a single organ but as a *congeries* of organs, each with a specific mental function. While earlier thinkers had posited some of these, Gall engaged in amplification or *copia,* divining from his observations a larger series of organs that correlated with various capacities.

Gall's collaborator, Spurzheim, took this approach even further after they ended their collaboration in 1812. Spurzheim codified the list of organs that he and Gall had developed earlier by using the rhetorical figure of *polyptoton.* The original list of organs came from Gall's observations of individuals: individuals who stood out for their specialized talents, individuals notable their idiosyncrasies, individuals who were imprisoned, individuals with mental illness, and so on. Based on these observations, Gall proposed categories such as "faculty of language" and "mechanical skill" (reflecting those with specialized talents) but also "carnivorousness," "sense of cunning," and "larceny" (representing the criminal classes he analyzed). Spurzheim invented words by adding new suffixes to existing words. He explains that he adds "-iveness" to refer to propensities, such as "amativeness" or "alimentiveness," and "-ousness" to indicate sentiments, such as "cautiousness" or "covetiveness."[31] The strangeness of this list reminds us that any qualities chosen for neuroscientific study are an artifact of the researchers' context and their dominant paradigms. While Gall had developed twenty-seven such faculties, Spurzheim added eight more for a total of thirty-five.

---

29. Gall, *Sur l'origine des qualités morales et des facultés intellectuelles de l'homme,* vi, my translation.

30. Bonnet, *La palingénésie philosophique,* 9.288; Cabanis, *Rapports du physique et du moral de l'homme,* 127–28, my translation.

31. Spurzheim, *Outlines of Phrenology,* 128.

Phrenology was, from the start, controversial. In France, in particular, Gall and Spurzheim received only a "moderately favourable" review from the National Institute of Arts and Sciences in 1808. Led by Georges Cuvier, the review did not acknowledge the significance of "cranial bumps" or the theory of localization.[32] Others called attention to the racist assumptions that undergirded phrenology, which positioned European heads as models of perfection and marked other races as inferior by virtue of their supposedly inferior physiognomy.[33] Yet despite skepticism about the field of phrenology writ large, researchers who engaged in anatomical dissections continued to use the metaphor of "organs" to describe the function of the brain.

The debate over localization of brain function did not die out as phrenology shifted from scientific inquiry to popular fad but instead shifted to questioning whether localization of brain functions could be determined not by looking externally, at bumps on the skull, but internally, at the brain's convolutions. Researchers continued to search for "organs" but began to use anatomical dissections as their method. While the metaphor of the "organ" remained, by the second half of the nineteenth century another metaphor surfaced, that of the "seat." In 1861 Paul Broca published "Remarks on the Seat of the Faculty of Articulated Language," positing a link between damage to the frontal lobes and loss of the ability to speak. His use of the term *seat* (or *siège* in French) parallels earlier metaphors that constitute the brain both as a system of government and as topography: the term *seat* might refer, metaphorically, to the seat of government (a geographic usage) or to a seat in a government or religious body. A "seat" in this sense is a site of authority (religious, political, and or administrative); the idea is that a site in the brain exercises authority over a particular function. Each faculty is a ruler over its own little cerebral domain—here, we see Platonic idea of the brain as city-state percolating up again, but the particular use of the term *seat* has a different rhetorical history.

In earlier medical discourse, it was common to say that the seat of a sensation, emotion, or disease lay in a particular organ. The theory of humoralism advanced by ancient Greek and Roman physicians assumed that temperament and health were based on four bodily fluids, each pertaining to a different organ (the liver, gall bladder, spleen, and brain or lungs). One of the earliest English usages of the term *seat*, in this context, is in John Gower's poem *Confessio Amantis*, where he states, of the yellow bile, or *chole*, that "his proper sete / Hath in the galle, where he dwelleth."[34] Even when the theory of humors was dropped, the metaphor of the seat persisted.

32. Simpson, 477–78.
33. Hamilton, "'Am I Not a Man and a Brother?,'" 176.
34. Gower, *Confessio Amantis*, 100.

In particular, the head or brain came to be described as the seat of a variety of things—wit, thought, consciousness, and so on. By 1777 Joseph Priestly would argue that, "Had we formed a judgement concerning the necessary seat of thought . . . we could not but have concluded, that in man it is a property of the *nervous system,* or rather of the *brain.*"[35] Similarly, Descartes described the brain as "the organ, or seat, *of Common sense, imagination, and Memory.*"[36] (Note that for Descartes, the terms *organ* and *seat* were interchangeable.)

The metaphor of the seat helped to fix mental faculties in a particular spot in the brain. It had the advantage, over the term organ, of seeming more precise and of dispensing with the comparison of the brain to a body. After all, organs in the body are fairly discrete, with relatively clear borders, but regions of the brain are not always so neatly divided. In addition, the metaphor of the seat allowed for a topographical way of thinking about the brain. Just as divisions of land into counties is invisible, with imaginary boundaries between counties, so too are the divisions of brain regions sometimes invisible aside from the more easily delineated lobes. Yet it is not the case that the metaphor of the organ simply disappeared and that of the seat emerged. Indeed, these metaphors intermingled in Broca's work and in later work.

As the nineteenth century progressed, focus shifted toward the nervous system as a whole, and this shift happened even before Santiago Ramón y Cajal famously identified the structure and function of nerve cells in the late 1880s. Indeed, it seemed to coincide with the emergence of other systems that featured prominently in nineteenth-century societies, such as the railway system. The Scottish psychologist and rhetorician Alexander Bain featured railway metaphors prominently in *The Senses and the Intellect* (first published in 1855), as in his description of the nervous system, where he noted that the nerves "form the medium of communication" between the body's extremities and its "centre,"[37] that the pons could be considered a "grand junction" between the medulla oblongata and the spinal cord,[38] and that the gray matter in the brain represents a "terminus."[39] These train metaphors offer an example of how material, economic, and technological forces constitute our understandings of scientific phenomena.

For Bain it was obvious that "the brain is the organ of the mind,"[40] but it remained to be determined whether and how the various sensory "centres"

---

35. Priestly, *Disquisitions Relating to Matter and Spirit,* 27.
36. Descartes, *L'homme,* 103, my translation.
37. Bain, *The Senses and the Intellect* (1855), 12.
38. Bain, *The Senses and the Intellect* (1855), 19.
39. Bain, *The Senses and the Intellect* (1855), 28.
40. Bain, *The Senses and the Intellect* (1855), 10.

were arranged in a network or system. These metaphors led Bain to shift away from localization and toward a more holistic understanding of the embodied mind. Bain argued, for instance, that an understanding of the nervous system means that "we cannot separate the centres [of the brain] from the other organs of the body that originate or receive the nerve stimulation. The organ of the mind is not the brain by itself: it is the brain, nerves, muscles, organs of sense and viscera."⁴¹ For Bain, it was the "transmission of influence along the nerve fibres from place to place" that "seems the very essence of cerebral action," and that transmission should not be confined to the brain itself but understood as a matter of "communications between the brain and the rest of the body."⁴² By carrying out the implications of his guiding railway metaphor, Bain developed a different understanding of the brain and its relationship to the body: this particular metaphor returned focus to the entire body (as in some of the earlier ancient Greek understandings) and not just to the brain.

This "extended body" understanding of cognition was, alas, restricted once again in the early twentieth century. In this period, the term *neural correlates* appears. It is found, for instance, in Edward Lee Thorndike's *The Elements of Psychology,* published in 1913: "The action in the nervous system which is connected with any mental state is called the *physiological or neural or nervous basis or correlate* of that mental state."⁴³ While Thorndike lists many possible terms here, offering two alternatives for "neural" and one alternative for "correlate," he ultimately settles on "neural correlate" as his preferred term. Since then, the phrase "neural correlates of x" has become the dominant construction used to link mental concepts with their material locations in the brain. "Neural correlate" is particularly interesting here because it links the physiological to the psychological. That is, it is used especially to refer to the neural correlates *of something,* a sensation, emotion, or mental operation. This term does something that the metaphors of the brain as a series of organs, seats, and centers, did not: it links biology to psychology.

Indeed, the term *neural correlate* emerges precisely when researchers were seeking to link physiology and psychology more systematically. In 1905 William McDougall made an argument for the importance of linking the two areas of study in *Physiological Psychology.* McDougall argued that psychologists should have a good understanding of physiology because it was probable that mental states (the focus of psychologists) corresponded to nervous states.

41. Bain, *The Senses and the Intellect* (1868), 52.
42. Bain, *The Senses and the Intellect* (1868), 52–53.
43. Thorndike, *The Elements of Psychology,* 170; emphasis added. The term *neural correlate* is an example of what Jeanne Fahnestock refers to as a term generated out of polyptoton, or repetition of a root term in a variety of forms.

McDougall used the term "neural correlate" to refer to those states that corresponded to the feelings.[44] McDougall builds his argument for the existence of "neural correlates" by referring to findings that he considered generally accepted. He declares that it was now accepted that "the destruction of certain parts of the brain always results in the loss of certain faculties or puts an end to certain kinds of experience," an oblique reference to the case of Broca's patient, Tan.[45]

For McDougall, the goal of physiological psychology was to "discover the relations of the two kinds of processes, the nervous and the psychical," in order to lay the foundations for a comprehensive understanding of human conduct.[46] He even offers an early neuropolicy argument, claiming that eventually this complete knowledge would move psychology beyond "purely academic study" by providing "the only sound theoretical basis for the art of the teacher."[47] While arguing strongly for the importance of psychological methods, including introspection, McDougall also held out for the importance of physiological study, with the idea that combining the two would provide a more complete account of human behavior than either one on its own.

The term *neural correlates* served a key function in this argument, a sort of rhetorical glue holding the psychological and the physiological together. The term *neural correlate* functions similarly to the term *gene* in genetics. In her rhetorical genealogy of genetics, Elizabeth Parthenia Shea shows how the term *gene* appeared long before the structure of DNA was discovered. Instead, *gene* served as a placeholder for a hypothesized function, allowing researchers to theorize about the mechanism behind inheritance and to establish its discovery as a key goal for research. It also epitomized a line of argument that a predictable, material function underlay the principles of Mendelian inheritance.[48] Similarly, the term *neural correlate* serves rhetorically as a placeholder and as an epitomized argument in psychological discourse. As a placeholder, *neural correlate* specifies the yet-to-be-identified material, physiological processes that correspond with a mental state. As an epitomized argument, *neural correlate* sums up the entire line of reasoning that McDougall and others advanced, namely that mental states were not the property of an immaterial soul, but that they had a material reality in the nervous system and especially

---

44. McDougall, *Physiological Psychology*, 81.
45. McDougall, 4.
46. McDougall, 5.
47. McDougall, 12.
48. Shea, *How the Gene Got Its Groove*, 66.

the brain. The term *neural correlates* supports an argument for neurorealism, for the material existence of mental concepts in the brain.

While the term *neural correlate* persists in the present, over the course of the twentieth century other controlling metaphors emerged that have also shaped current research practice. For one, the brain began to be associated with the computer, and vice versa. In fact, the first appearance of the brain-as-computer metaphor appeared in the reverse, with the computer being referred to as a brain. Shortly after World War II, wartime developments were revealed to the public. In 1945, for instance, a *New York Times* article reported on a flight simulator that used an "electronic brain" composed of 220 vacuum tubes; similar "electronic brains" were described that could be used by airlines to check the availability of seats on flights or to accurately cut and shape propellers for planes.[49] So many "electronic brains" were reported in the popular media that a backlash soon followed. In 1947 Donald Barr complained in the *New York Times* that "there is altogether too much mischievous gibberish of the miracle drug and electronic brain type" in scientific popularizations;[50] in 1949, also in the *New York Times,* an editorial noted that "so many 'electronic brains' have made their appearance in the last ten years that this department finds it hard to keep up with them."[51]

While some were confident that "electronic brains" could parallel or even exceed the operations of the human mind, the term also raised concerns. In the *British Medical Journal,* for instance, a 1946 editorial admitted that, while the "electronic brain" could perform "feats of discrimination and subsequent action," it could not be said to engage in "thought."[52] Similarly, an article in *The Lancet* from 1946 noted that, in humans, "memory involves the presentation of the past to consciousness, and to apply this term to a calculating machine except by analogy is to confuse the material basis of memory with memory itself."[53]

Despite these reservations, soon the suggestion that the computer was a brain led to the reverse—the suggestion that the brain might be a computer. In 1949, for instance, an article in the *British Medical Journal* described computers as "'electronic' or 'mechanical brains'" but noted that this terminology has led "to the suggestion that the human brain is itself a network of units of this

---

49. "Air Travel in West Increases Sharply," "Electronic Control Speeds Production of Ship Propellers."
50. Barr, "A Decade of Miracles, Once Over Lightly," br18.
51. "Electronic Brain Does Research."
52. "Electronic Brain."
53. "The Electronic and the Human Brain," 795.

general type."[54] The idea of the brain-as-computer/computer-as-brain was, for some, more than a mere metaphor. According to W. Grey Walter, computers actually did reflect the structure of the human brain: "The engineers who have designed our great computing machines adopted this system without realizing that they were copying their own brains. (The popular term electronic brain is not so very fanciful.)"[55] Similarly, in "Mechanical Concept of Mind," written in 1953, Michael Scriven notes that computers "may be entirely similar in construction to a human brain."[56] In inventing the computer, these thinkers believed, scientists had duplicated the structure of the brain.

Of course, the brain-as-computer/computer-as-brain figure reaches its apogee in cybernetics, a term originated in the late 1940s by Norbert Wiener. An interdisciplinary collaboration that grew out of World War II, cybernetics sought to connect research in fields such as computing, biology, engineering, and mathematics. In *Cybernetics,* Wiener outlined how the field focused, in particular, on connections between computing machines and the nervous system. In the process, he generated a series of metaphors that transferred meaning from the former (and specifically the realm of communications engineering) to the latter, such as "feed-back," "message," "prediction mechanisms," "background noise," and "information."[57] The premise was that the basic principles of "communication, control, and statistical mechanics" would be the same regardless of whether one was considering machines or human tissue.[58] In this way, Wiener evokes a type of neurorealism, suggesting that the problems related to the function of synapses were "after all switching problems" and that "the synapse *is nothing but* a mechanism for determining whether a certain combination of output from other selected elements will or will not act as an adequate stimulus for the discharge of the next element."[59] Cybernetics functioned on a premise that almost exceeds mere metaphor or analogy: for Wiener and his colleagues, the nervous system so closely parallels a computer that it may nearly be treated as one.

This metaphor not only shaped how cybernetics researchers understood the brain but also constrained how they understood psychological concepts, which had to be constituted as a form of information with an input and out-

54. Newman, "Electric Automatic Computing Machines," 1133.

55. Walter, "An Imitation of Life," 43.

56. Scriven, "The Mechanical Concept of Mind," 236.

57. Wiener, *Cybernetics; or, Control and Communication in the Animal and the Machine,* 13–18.

58. Wiener, 19.

59. Wiener, 21–22, emphasis added.

put.[60] As N. Katherine Hayles explains, this understanding of the brain (and of computers) depends on a reification of information in a way that strips it of any context, making it "a mathematical quantity weightless as sunshine, moving in a rarefied realm of pure probability, not tied down to bodies or material instantiations."[61] This view of information was not the only one being considered at the time, but the alternative—a model of information that understood the importance of context and embodiment—was less practicable in an engineering context because it was more difficult to quantify and could not easily be rendered universal. We see a similar challenge in social and cognitive neuroscience, where embodiment and context continue to be set aside given technical constraints, especially in the case of fMRI.

The analogy computer-as-brain/brain-as-computer led to one further metaphor that deserves attention here: that of the circuit or network. As Turing put it, "Many parts of man's brain are definite nerve circuits required for definite purposes. Examples of these are the 'centres' which control respiration, sneezing, following moving objects with the eyes, etc.: all the reflexes proper . . . are due to the activities of these definite structures in the brain."[62] Here, we see the emerging network or circuit metaphor working alongside the earlier metaphor of the center as well as the architectural metaphor of the "structure." This set of mixed metaphors positions earlier, locationist ways of thinking of the brain (seats, organs, circuits) alongside newer, connectionist or network-driven ways of thinking (networks, circuits, etc.).[63] This pattern continues in contemporary social and cognitive neuroscience that uses fMRI, which not only has inherited this mixed set of metaphors but seems to materialize those metaphors in the very form of functional magnetic resonance imaging, which translates blood oxygenation activity in tiny three-dimensional units of the brain, or voxels, into information (data) that is then translated into a colorful image.

Locationist metaphors are not without detractors, however. According to some researchers, the localization paradigm operates on a false assumption: it ignores the potential for multiple brain areas to be invoked in a single function, and for multiple functions to occur within a specific brain region. Psychologist William Uttal argues in his book *The New Phrenology* that "every

---

60. Fahnestock, "Rhetoric in the Age of Cognitive Science," 161.
61. Hayles, *How We Became Posthuman*, 56. Consult Hayles for a deeper history of cybernetics and its function in forming an equivalence between humans and machines. See also Pruchnic, "Neurorhetorics: Cybernetics, Psychotropics, and the Materiality of Persuasion."
62. Turing and Copeland, *The Essential Turing*, 423.
63. Notably, however, Turing believed that much of the brain was largely indeterminate in function, rather than being specialized into discrete tasks or localized actions (Turing and Copeland, 424).

study of the localization issue and every theory about it is premised on the idea that variations in the psychological domain are in some very direct way related to variations in the neurological domain."[64] Instead, he argues, "it seems far more likely that the mind, consciousness, or self-awareness represent the merging or binding of many different underlying processes and mechanisms into an integrated singular experience."[65] Similarly, philosopher Alva Noë insists that consciousness (by which he means thought, awareness, and "being in the world,") has no locus in the brain; instead, "it is something that we do, actively, in our dynamic interaction with the world around us."[66] And philosopher Andy Clark argues that human cognition "leaks out into body and world," extending beyond the brain itself,[67] even if the "the material vehicles of conscious experience" do not actually extend outside the brain and nervous system.[68] These researchers prefer metaphors of extension, interaction, and fluidity, understanding the brain in a way that allows for interaction of brain areas, or as a part of a broader system that includes our bodies and the world around us.

However, connectivist and embodied metaphors still lead to the same fundamental problem as all metaphors. For Burke, all metaphors necessarily lead to reductive thinking because they are inherently perspectival. In the case of the human sciences, Burke insists that "any attempt to deal with human relationships after the analogy of naturalistic correlations becomes necessarily the *reduction* of some higher or more complex realm of being to the terms of a lower or less complex realm of being."[69] Attempting to understand human nature, in all its complexity, as a function of the brain necessarily partakes of this reductiveness, but so do other perspectives relying on different metaphors. This question then becomes whether we can still learn from those reductions while recognizing them as such, with a critical awareness of their limitations.

HOW NEUROSCIENTISTS STUDY DECISION-MAKING

Researchers are always relying on metaphors as the guiding principle or paradigm for research or as models to direct their lines of inquiry, and those

---

64. Uttal, *The New Phrenology*, 4.
65. Uttal, 6.
66. Noë, *Out of Our Heads*, 24.
67. Clark, *Supersizing the Mind*, xxviii.
68. Clark, "Dreaming the Whole Cat," 753.
69. Burke, "Four Master Tropes," 424.

metaphors can lead to productive findings. Yet, they also lead researchers to ignore other possible avenues of research, other findings that would be illuminated given a different trope. The question is not whether the metaphor is permissible but whether it is useful—does it serve the scientists' purposes? Is it appropriate and useful? What other metaphors might offer new perspectives?

To investigate this question, consider how neuroscientists currently model decision-making—of which buying behavior, to return to the opening anecdote for this chapter, is an example. Currently, neuroscientists investigate decision-making within a framework commonly referred to as "neuroeconomics." While one might expect the term to refer specifically to decision-making about economic decisions, it tends to treat all areas of decision-making within an economic framework. Within this framework, the brain is figured as a computer or mechanism for calculating risk, benefit, value, and reward. These terms clearly reflect a capitalistic model. As Michael Platt and Scott Huettel note, decision-making studies are almost always done using monetary rewards (at least when humans are the research subjects).[70] Typically, these rewards are provided in gambling tasks that require participants to make decisions based on maximizing profit or on evaluating "trade-offs between economic parameters, such as a choice between one outcome with higher expected value and another with lower risk."[71] The assumptions embedded here are seldom questioned by neuroscientists: it is simply assumed that human subjects value money and that money can represent a universal form of reward. The idea of maximizing profit is also taken as natural. Then, this model of decision-making about economic tasks is taken as representative of decision-making writ large.

Notably, this paradigm even transfers over to animals. Platt and Huettel describe a study featuring birds that were given food as a reward, but they employ monetary metaphors to describe the study: the birds are described as calculating a "rate of gain," as "gambling on the risky option," as "hitting the jackpot," or as engaging in a "subjective valuation of rewards" in order to require "resources."[72] In other words, not just the language but the economic way of *thinking* about decisions is so naturalized that researchers can apply it to birds without seeming to notice. These metaphors then materialize in study designs that often take the form of a lottery or gambling task with rewards that are either symbolic (for instance, a monetary amount displayed on a screen) or literal (for instance, actual money or food).

---

70. Platt and Huettel, 399.
71. Platt and Huettel, 400.
72. Platt and Huettel, 399.

The language of neurorealism concretizes and naturalizes this way of thinking. Take the following paragraph from Platt and Huettel:

> How does the brain deal with uncertainty? Are there distinct regions that process different forms of uncertainty? What are the contributions of brain systems for reward, executive control and other processes? The complexity of human decision making poses challenges for parsing its neural mechanisms. Even seemingly simple decisions may involve a host of neural processes.[73]

The first sentence positions the brain as an active agent that "deals" with uncertainty. This construct is a marker of neurorealism because it positions the brain as an active, separate agent in its own right. The following sentences employ several different neurorealist metaphors—that of regions (which we have seen was prevalent in earlier periods), that of "systems," that of "mechanisms," and that of "processes." Together, these metaphors evoke the dominant framework of the brain as computer but suggest legacies of earlier metaphors of the brain as map (with regions and centers) and the brain as machine (with mechanisms and functions).

In these paradigms, rationality and logic are taken as, if not the norm, the presumed ideal of decision-making. We can find an alternative to this view in rhetorical traditions that have traditionally understood decision-making in less rigid terms. Rhetorical decision-making focuses not on the certain but on the contingent, not only on fact but on opinion. As Gorgias puts it in the *Encomium of Helen,* "if all men on all subjects had both memory of things past and awareness of things present and foreknowledge of the future, speech would not be similarly similar, since as things are now it is not easy for them to recall the past nor to consider the present nor to predict the future. So that on most subjects most men take opinion as counselor to their soul."[74] For rhetorical scholars, ambiguity and uncertainty are not a bug but a feature (to borrow the computational metaphor). That is, a rhetorical approach assumes that most situations are fuzzy, requiring people to deliberate and make decisions in cases where information is incomplete, different values are at stake, or the needs of different stakeholders may be in conflict. In the next section I will suggest how neuroscientists could benefit from a rhetorical perspective of decision-making and deliberation that provides an alternative to a rationalistic, computational model.

---

73. Platt and Huettel, 399.
74. Gorgias, *Encomium of Helen,* 52.

A NEURORHETORICAL APPROACH TO NEUROREALISM

Neurorealist metaphors are important because they shape how we understand the target (here, decision-making) and the brain itself. They connect. In our ravelings, it is important to use the lens of neurorealism (and neuroessentialism, as I will show in the next chapter) to critically interrogate social and cognitive neuroscience theories. One way to begin is by identifying central metaphors. Each metaphor favors one understanding of the brain over others, potentially misleading us in the process. Rhetoricians seeking to work with the neurosciences have the luxury of choosing among, critiquing, and adding to those theories and the metaphors that inform them. For neuroscientists, though, the situation is more complex because neorealist metaphors are deeply embedded in the discipline. At the very least, then, it is important to acknowledge the metaphors that guide a line of inquiry; at best, one might also ask what other metaphors are available. A neurorhetorical approach to neurorealism, then, might ask the following questions.

1. What are the guiding metaphors?

In neuroscientific studies of decision-making, multiple metaphors are in play. The metaphor of the computer is particularly dominant here because, just as the neuroeconomic framework posits a binary system of positive or negative gain, so too does a computational system posit a binaristic system of positive and negative signals. The two frameworks work in concert. Ultimately, this framework positions the brain as a calculator of risk and reward—the human equivalent of the Wall Street hedge-fund manager. For neuroeconomists, the brain *really is* a computer or calculator: this is not just a textual feature or even a rhetorical strategy but a disciplinary paradigm through which they study the brain.

2. For what do these metaphors select and from what do they deflect our attention?

Within the brain-as-risk-calculator paradigm, it is surprising when people do not act in a purely rational manner. After all, within an economic system people (and animals) presumably act in ways that lead to the greatest profit. The fact that humans frequently act irrationally has been a source of consternation and puzzlement in that discipline. Thus, this metaphor directs our attention toward "rational" behavior that seeks to maximize profit and away from altruistic behavior, which is seen as "irrational." Clearly, such a brain-as-risk-calculator metaphor squares with a Western, capitalist outlook, but would sit less well with a framework that values care for others, nurturance, or altruism.

3. What other metaphors are possible?

While the brain-as-risk-calculator draws our attention to some aspects of decision-making, it overlooks others. New metaphors might offer different lines of inquiry. For instance, different metaphors might help researcher consider altruism not as a bug in a profit-calculating machine but as a feature of human cognition. What would it mean to think of the brain as a triage unit, assessing needs and responding accordingly? In a triage unit of a hospital, decision-making still depends on ranking and weighing needs, but it focuses on attending to the needs of others, not just the self.

The brain-as-risk-calculator also tends to either ignore emotions, to position them as "irrational," or to position them as "somatic markers" that improve the same kind of risk-benefit decision-making that goes on in the dominant framework.[75] Such is not the case in rhetoric. The earliest rhetorical theories from Gorgias onward have all stressed the significance of the emotions and the body, and one of our earliest texts in fact explores a situation where someone seemed to have been acting irrationally. In the *Encomium of Helen,* Gorgias tries out several lines of reasoning in order to exonerate Helen, who seems to have thrown rationality aside when she went off with Paris. In his speculation about the power of language, Gorgias clearly describes the embodied feelings that speech evokes—its ability to "stop fear and banish grief and create joy and nurture pity" as well as its effects on the body. He describes the "fearful shuddering and tearful pity and grievous longing" that come upon listeners to an oration. Gorgias also describes how similar embodied responses can occur through visual stimuli, as in the case when "belligerents in war buckle on their warlike accouterments of bronze and steel, some designed for defense, others for offense, if the sight sees this, immediately it is alarmed and it alarms the soul, so that often men flee, panic-stricken, from future danger as though it were present."[76] Thus, rather than positioning emotion as something that adds to reason, Gorgias acknowledges the overwhelming, sometimes irrational power of emotion. This perspective might lead us to develop new metaphors that acknowledge the sensitivity of the brain to emotions and affects.

---

75. Damasio, "The Somatic Marker Hypothesis."
76. Gorgias, 53.

4. How might different metaphors open up different lines of inquiry?

I'm not the only one to notice that the brain-as-computer metaphor, and its attendant assumptions, limit the range of inquiry. In an article that circulated widely online in 2016, Robert Epstein argued that the brain-as-computer metaphor misguides researchers for many of the reasons I've identified here, along with others.[77] For one, Epstein claims, this metaphor leads researchers to assume that the brain *stores* information somewhere, the way that computers do, rather than *responding* and *changing* as a result of new experiences or stimuli. Yet, when Epstein challenged researchers to identify different metaphors, no one could think of any alternatives. Epstein argues that, instead, researchers should pursue a "metaphor-free theory of intelligence," but from a rhetorical perspective, such a theory would be difficult to talk about, let alone pursue. After all, from a rhetorical perspective *all language* is metaphorical or, at least, tropological.

Here is where humanists might come in: we are good at thinking heuristically about metaphors. The goal of such an inquiry might not be to identify a single, dominant metaphor to guide all future research but to generate multiple alternatives that could open up new lines of inquiry. What would it mean to think of the brain as a sponge? As an ecosystem? As a rhizome? As an organism? Or as a quantum field, as Daphne Muller has argued?[78]

Regardless of which metaphors are taken as generative, the goal is to *recognize* that metaphors are metaphors. In other words, it might be better to prefer simile over metaphor, to say that the brain is "like a sponge" or "like an ecosystem" or "like a computer"; to prefer partial similarities to total resemblance. Then, one might specify in what ways the brain is like something *and* in what ways it is unlike it.

Thus far, then, I have shown how metaphors are central to the problem of neurorealism. Yet neurorealism also depends on synecdoche, or the ability of a part of something to stand in for the whole. Neurorealism works hand in hand with synecdochal reductions in meaning for whatever is being measured. This reduction, I will argue, constitutes neuroessentialism, another key rhetorical operation upon which neuroscientific inquiry depends.

---

77. Epstein, "The Empty Brain."
78. Muller, "Your Brain Isn't a Computer—It's a Quantum Field."

CHAPTER 2

===

Neuroessentialism

A FREESTYLE RAPPER lies in an fMRI machine. He is given an instrumental musical track and asked to freestyle along with it while neuroscientists record his brain activity. In another lab, a dancer lies immobile, just like the rapper, only she is asked to choreograph the craziest dance she can think of—in her head. Her brain activity is being measured too. And somewhere else, a study participant—not a creative writer—lies in an fMRI machine and is asked to compose a story. All these studies ask participants to do different things. But they are all cast as studies of "creativity." How is it that these different activities, in different modes (oral, embodied, and written) all get to count for the same thing?

To research a concept such as creativity using most neuroscientific principles and methodologies, it is necessary to argue it into place as real, as existing within the brain, using the resources of neurorealism described in the previous chapter. But researchers must also define what they mean by creativity. To do so, they draw on the rhetorical figure of synecdoche, reducing the wide range of possible understandings of a quality like creativity and the range of ways that it may be materially enacted, performed, measured, into something that can be easily and neatly implemented in an fMRI study. Then, researchers can go on to identify its "neural correlates" or to locate it in a "center" or "network," to study its functions among different groups (artists, authors, musicians), and even to suggest potential applications of their research for policy.

In this chapter, I will demonstrate how neurorealism functions rhetorically in the experimental design of fMRI neuroscience studies. Next, I will ravel back, providing a genealogy of neuroessentialism. Then, I will examine instances of creativity's enactment in studies such as the one of freestyle rappers, suggesting how creativity is constituted within the experimental setting in ways that essentialize it as a particular type of improvisatory act. Finally, I will suggest how a rhetorical perspective might enrich those perspectives in the case of creativity, an area that has long been of interest to rhetoricians as well as psychologists and neuroscientists.

HOW NEUROESSENTIALISM WORKS

Neuroessentialist claims operate at the second stasis, that of definition. Racine, Bar-Ilan, and Illes define neuroessentialism as arguments that "equat[e] subjectivity and personal identity to the brain."[1] Racine, Bar-Ilan, and Illes worry mainly about the kind of essentialism that applies to people. In the case of creativity, researchers often focus in on people who are already deemed to be creative—such as dancers, rappers, or writers. Then, by analyzing their brains, they can claim to have identified key features of creativity. The study design enacts essentialism both by equating creativity with certain kinds of people (creative types) and also by equating creativity only with certain kinds of acts, namely improvisation.

Racine, Bar-Ilan, and Illes pin the problem on popular news accounts; however, I argue that neuroessentialism is a much deeper methodological problem. Before mapping any mental concept onto the brain, it must be defined in such a way that it can be operationalized. In fMRI studies, this means choosing a definition that fits the limitations of the technology. Individuals cannot be observed in a naturalistic setting while in an fMRI machine. Dancers cannot actually dance; artists cannot actually paint or sculpt. They must lie still, avoiding head movements, and can usually only respond by clicking a response button. They can be shown images, listen to sounds, and even experience smells. Their heart rates, skin conductance, or respiration may be measured. They can make choices, clicking the button when they see a certain shape or color. And they may do mental problems or tests, they may think about certain things and not others. But enacting a more complex social situation is impossible. Only representations or simulations of a social situation are possible. Thus, to define a mental concept like creativity or pleasure

---

1. Racine, Bar-Ilan, and Illes, "fMRI in the Public Eye," 160.

or love and operationalize it in an fMRI inevitably revolves essentialism. The concept in question must be boiled down to something measurable, reducing it to a concrete action or prompt that is then posited as the "essence" of that concept. They must also be contrasted with either a neutral state or something that can be opposed to whatever is being studied. So, "creativity" must be measured against something else, like "rote repetition"; pleasure to pain, and so on.

Neuroessentialism claims are fundamentally reductive. As Kenneth Burke explained, "any attempt to deal with human relationships after the analogy of naturalistic correlations becomes necessarily the *reduction* of some higher or more complex realm of being to the terms of a lower or less complex realm of being."[2] As the *sine qua non* of "scientific realism," reduction often occurs in a movement from quality to quantity.[3] Rather than measuring subjective responses to pleasure, for instance, that response comes to be measured by the number of times a rat presses a lever, or by the concentration of oxygenated blood in a particular brain region.

Neuroessentialist claims also frequently work by articulating a quality alongside a particular region in the brain (thereby linking it to a neurorealist metaphor such as the ones described in the previous chapter). So, instead of talking simply about creativity, rhetors can talk about creativity as grounded in, say, the prefrontal cortex, thereby lending their claims extra scientific credibility. Neuroessentialism connects a quality with a location in the brain, giving that quality a realness and lending the rhetor scientific credibility. Thus, it hinges on the first stasis of fact as well as definition, and therefore tends to embed within it neurorealist claims:

- Neurorealist claim: creativity exists in the brain
- Neuroessentialist claim: creativity is the ability to improvise (say, a dance, rap, or written composition)

These claims are important because they simplify concepts and regions that are inevitably complex and fuzzy. They serve inventive purposes for scientists, who must use them to generate experimental protocols, articles, and funding. Yet neuroessentialism can be problematic because it flattens out the concept in question (in this case, creativity), simplifying the complex training, habits, material tools and practices, cultural conditions, and cognitive functions that constitute it.

---

2. Burke, "Four Master Tropes," 424.
3. Burke, "Four Master Tropes," 429.

Another way of understanding neuroessentialism would be to say that neuroscience studies rely upon what rhetorician Edward Schiappa would refer to as "facts of essence." Their approach assumes that mental concepts have "real" or "true" definitions—those uncovered by neuroscientists. According to Schiappa, essentialist definitions "depend on metaphysical absolutism: the belief that things have independent, 'objective' structures or essences that are knowable 'in themselves.'"[4] One of the primary challenges facing the social and cognitive neurosciences is the problem of metaphysical absolutism, and it is a problem that has no clear solution because it is deeply embedded in the research practices, instruments, methods, and histories of the field.

For one, as Nikolas Rose and Joelle Abi-Rached have argued, social neuroscience has established much of its territory by taking on conceptions of personhood that had already been colonized by psychology.[5] Terms such as *creativity* have been studied by psychologists, who have attempted to define them, operationalize them, and measure them through a number of psychometric tests, experiments, and procedures. Thus, social and cognitive neuroscientists rely on extant definitions (which are often themselves essentialist) and sometimes existing metrics (scales and questionnaires) and must simply develop ways of enacting (or operationalizing) them in an fMRI setting.

Neuroessentialist claims also mask the fact that the connections between brain concepts and regions are understood by scientists to be correlations or associations between a concept and region, not necessarily cause-effect relationships. For instance, while scientific studies may show some correlates between depression and certain patterns of neural activation, it is reductive to say that those patterns of neural activation constitute depression. As Alva Noë puts it:

> There are neural signatures of depression. Direct action on the brain—in the form of drug therapy—can influence depression. But in another sense, it is obviously not true. It is simply impossible to understand why people get depressed—or why this individual here and now is depressed—in neural terms alone. Depression happens to living people with real life histories facing real life events, and it happens not only against the background of these individual histories but also against the background of the phylogenetic history of the species.[6]

---

4. Schiappa, *Defining Reality*, 35–36.
5. Rose and Abi-Rached, *Neuro*, 9.
6. Noë, *Out of Our Heads*, xii.

We might consider the materiality of depression as well by asking, for instance, how the built environment, our interactions with technology, animals, places, and things may *enact* depression, or how depression (which is, after all, a psychiatric diagnosis) is enacted in a psychiatrist's office versus the pharmacy or the office of accessibility at a university. In what follows, I will use a set of two examples to illustrate how a rhetorical perspective, specifically, can enrich neuroscience studies of creativity. While Noë calls for a deeper sense of individual and species-specific histories, a rhetorical perspective calls attention to the kinds of trained, habituated, embodied responses (what Burke might called "trained incapacities" but we might also refer to as "trained capacities") that are overlooked when creativity is treated as simply a quality that one does or does not intrinsically possess.[7] First, though, it is important to understand how mental concepts, like creativity, have historically been identified in psychological thought, as well as how, in the twentieth century, researchers have sought to measure them quantitatively.

A RHETORICAL HISTORY OF NEUROESSENTIALISM

While philosophers have long sought to understand the human mind, the idea that this could be done by measuring mental faculties with scientific, quantitative precision is a relatively recent one. Previously, mental concepts were addressed qualitatively, through introspection and philosophical analysis. These introspective discussions usually led to theories that recognized the multiplicity and variability of these concepts. Plato devotes several of his dialogues to considering emotions, such as the *Phaedrus,* which meditates on love, and the *Philebus,* which weighs the relative merits of pleasure and wisdom. In *Philebus,* pleasure is understood to be multiple and contingent; here Socrates states that it "has taken all sorts of shapes that are in some sense unlike one another."[8] Similarly, Aristotle considered emotions as embedded in rhetorical situations. Anger, he writes, "may be defined as an impulse, accompanied by pain, to a conspicuous revenge for a conspicuous slight directed without justification towards what concerns oneself or towards what concerns one's friends."[9] Aristotle notes that anger differs depending on the situation. One might be angrier toward one's friends than toward other people, he notes, since "we feel that our friends ought to treat us well and not badly."[10]

---

7. Burke, *Permanence and Change,* 7.
8. Plato, *Philebus,* 12c4–8.
9. Aristotle, *Rhetoric,* Book II.2.1378b.
10. Aristotle, II.2.1379b.

In the *Rhetoric,* Aristotle provides an inventory of emotions, from anger to fear to shame, showing how each one depends on social contexts and situations. As Daniel Gross puts it, for Aristotle, "Anger is a deeply social passion provoked by perceived, unjustified slights, and it presupposes a public stage where social status is always insecure."[11] From this perspective, it would be difficult to measure or quantify an emotion such as anger, since doing so would involve sorting out the different kinds of anger that occur in different situations, for different reasons.

A contextual understanding continued in the Enlightenment, when philosophers developed taxonomies of mental faculties and emotions. John Locke's *An Essay Concerning Human Understanding* (1689) presents a taxonomy of mental concepts or operations, which he calls "ideas." These include those provided by the senses, as well as those achieved through mental operations, or "reflection." This last category includes "ideas" such as "perception, thinking, doubting, believing, reasoning, knowing, willing, and all the different actings of our own minds."[12] For each of these "ideas," Locke stresses that clear definitions and descriptions elude us, that they can be known only through experience. In chapter 20, "Of Modes of Pleasure and Pain," Locke outlines several mental concepts: pleasure and pain, good and evil, love and hatred, desire, joy, sorrow, hope, fear, despair, anger, and envy. Like Aristotle before him, he defines each one by the contexts in which they arise. Envy, for instance, "is an uneasiness of the mind, caused by the consideration of a good we desire obtained by one we think should not have had it before us."[13] Locke reflected on the quality and experience of each sensation, not on its definition or measurement.

Another contextual approach to faculty psychology is well known to rhetoric scholars. In the *Philosophy of Rhetoric,* George Campbell considers four mental faculties relevant to the practice of rhetoric: reason, imagination, memory, and the passions. To move the passions, for instance, one must first "excite some desire or passion in the hearers," and then "satisfy their judgment that there is a connexion between the action to which he would persuade them, and the gratification of the desire or passion which he excites."[14] To do so, the rhetor must understand the circumstances in which different passions occur.

This brief history of mental concepts suggests a tendency to view emotions or mental qualities as connected to contexts, rather than as discrete things or

11. Gross, *The Secret History of Emotion,* 2.
12. Locke, *An Essay Concerning Human Understanding,* 123.
13. Locke, 306.
14. Campbell, *The Philosophy of Rhetoric,* 94.

cerebral locations. They might arise within different social situations, or, in Campbell's thought, be evoked through rhetorical acts. The very fact that they are considered qualities, *not quantities,* suggests why it would not have been thinkable to measure them.

In addition, in this philosophical trajectory, thinkers recognized that mental concepts are themselves rhetorical—that is, they are constituted through (and therefore inseparable from) language. Some philosophers question whether one can separate these mental concepts from the language we use to describe them at all. Scottish philosopher Thomas Reid makes this point in his *Essays on the Intellectual Powers of Man* (1785): "Almost all the words by which we express the operations of the mind are borrowed from material objects."[15] Reid gives the examples of "conceive" (from the Latin *concipere,* to hold in hand), "imagine" (from the Latin *imago,* or painting), and "deliberate" (from the Latin *libare,* to weigh).[16] Our words for emotions are, in that sense, themselves metaphors.

Psychologist T. R. Sarbin argues similarly that mental-state terms "have no existent referents, only vague verbal ones. We learn these terms before we can be analytical about their origins and about their unreliability of the sensory inputs that call out these symbols."[17] According to Sarbin, most mental-state terms emerged in Europe in the sixteenth century, when religious acceptance gave rise to a belief in a private, internal world. Before that, concepts we would now distinguish as internal, mental ones were considered externalized, attributed to forces such as changes in seasons, bodily humors, or invasion by "demons" or spirits. Thus, the terms we use for mental states have their origins in terms used for external states or things. For example, anxiety comes from the word *anguish,* which comes from the French word *anguisse,* which Sarbin describes as a painful, choking sensation in the throat. This sensation comes to be applied, metaphorically, to mental anxiety.[18] The terms we use for mental concepts thereby serve as shorthand for talking about complex, embodied experiences and emotions in more tangible terms.

The terms *creative* and *creativity* have a similar history of being understood, first, as external forces. According to the *Oxford English Dictionary,* the term was first used in English in specifically religious contexts. It was God, originally, who possessed creative power. For instance, in John Irland's *The Meroure of Wyssdome,* the author insists that creative power over the soul

---

15. Reid, *Essay on the Intellectual Powers of Man,* 237.

16. Reid, 237.

17. Sarbin, "Anxiety: Reification of a Metaphor," 632.

18. Sarbin, 633.

was granted to God alone, not humans.[19] By extension, the term began later to be applied to skill in what we would now call creative arts. It appears, for instance, in Richard Savage's 1729 poem, "The Wanderer," in the opening portion where the poet invokes the muses:

> Come CONTEMPLATION, whose unbounded Gaze,
> Swift in a Glance, the Course of Things, surveys;
> Who in Thy-self the various View can'st find
> Of Sea, Land, Air, and Heav'n and Human Kind;
> What Tides of Passion in the Bosom roll;
> What Thoughts debase, and what exalt the Soul;
> Whose Pencil paints, obsequious to they Will,
> All thou survey'st, with a creative Skill![20]

Similarly, the noun *creativity* emerges first with reference to God's unique powers. In Goerge Lawson's *Theo-Politica,* for instance, creativity describes powers not granted to humans, since God's creation of the Earth was "far different from the Acts of any Creature."[21] Only later was the term extended to a mortal, as in Adolphus William Ward's use of the term to describe Shakespeare's prowess in his *History of English Dramatic Literature.* Ward describes Shakespeare's poetry as "unrivalled," due to "the spontaneous flow of his poetic creativity."[22]

Thus, from the start of its usage in the English language, at least, creativity is associated with the divine—with an innate, supernatural (or at least extraordinary) ability, rather than as something that emerges out of training or practice. Yet, aside from God, even the most talented individuals seem to require training and some prior materials in order for their creativity to manifest. We know that Shakespeare, for instance, composed his works by drawing on available materials in the tradition of *imitatio.*[23] His historical tragedies, such as *Richard III* and *Henry VI,* draw on Holinshed's *Chronicles of England, Scotland, and Ireland,* whereas his *Julius Caesar* and *Titus Andronicus* draw upon Plutarch's *Life of the Noble Grecians.* This is not to diminish Shakespeare's skill but to qualify it. That is, Shakespeare (like any creative writer) worked by drawing on available materials, by reworking plots, fleshing out or developing new characters, and so on. Thus, his "creativity" emerges out of something

---

19. Irland, *The Meroure of Wyssdome,* 79.
20. Savage, *The Poems of Richard Savage,* 116.
21. Lawson, *Theo-Politica,* 57.
22. Ward, *A History of English Dramatic Literature to the Death of Queen Anne,* 506.
23. Newcombe, "Toward a Sustainable Source Study," 29.

akin to what Aristotle defines as rhetoric. While Aristotle defines rhetoric as the "faculty of observing, in any given case, the available means of persuasion," we might define the composing of creative works as the faculty of using, in any given case, the available means and materials for composing.[24] In neuroscience, however, concepts such as creativity are typically understood differently in order for them to be operationalized in fMRI studies; to do so, they draw on the tradition of psychometrics and usually provide very bare (if any) reference to this former way of understanding creativity as externally situated.

For neuroessentialism to emerge, mental concepts had to be not only linked to brain regions but reduced to measurable entities that could be usefully identified and tested using the kinds of neurological tools available. In large part, this relies on psychometrics, a discipline that emerged in the mid-twentieth century alongside innovations in statistical analysis, including multivariate statistics and factor analysis.[25]

Psychometrics provides what Jessica Mudry would call a discourse of quantification, a tendency that cuts across the life sciences and social sciences, but one that inevitably tropes that which it seeks to measure, often by conflating and confusing qualities with quantities.[26] A discourse of quantification, Mudry argues, "encourages certain interactions and eliminates others because only certain ideas, arguments, and topoi can operate within it."[27] In the case of psychological concepts, quantification excludes certain kinds of topoi (commonplaces) and arguments, as well as material practices, ephemeral sensations, culture, history, and individual differences. Yet, as Mudry explains, quantification is "the bedrock upon which much scientific communication happens and even acts to define a discipline as science."[28] Such is the case in psychology, where quantification lent an air of scientific authority to a discipline previously considered more closely aligned with philosophy and the humanities. According to JoAnne Brown, proponents of psychometric tests drew their quantitative language from the models of engineering and medicine, using that language to advance the status and authority of psychology as a discipline.[29]

The craze for measurement of psychological concepts (as opposed to brains themselves) began in the mid-nineteenth century, when psychologists began to employ scientific methods and techniques that were already accepted

24. Aristotle, *Rhetoric*, 24.
25. Merenda, "Measurements in the Future," 210.
26. Mudry, *Measured Meals*, 7, 9.
27. Mudry, 9.
28. Mudry, 5.
29. Brown, *The Definition of a Profession*, 5.

for use in other fields. In *Inventing Our Selves,* Nikolas Rose argues that statistics and the experiment were the two most significant of these techniques.[30] I would argue that a third technique, psychometric testing, was also crucial to psychology's ability to establish itself as a scientific discipline.

Psychometrics began in earnest in the early twentieth century, with the advent of the intelligence quotient, or IQ test, developed by French psychologists Alfred Binet and Theodore Simon, and later revised by American psychologist Lewis Terman. Intelligence tests were initially used to identify "grades" of individuals with intellectual disabilities, both in France and the US, and soon outstripped their original purpose. In his 1916 book, *The Measurement of Intelligence,* Lewis Terman identified several additional uses for intelligence testing, such as classifying the causes of delinquency, identifying "superior children" in schools, assigning new students to the appropriate grade, and determining the best type of vocational training for different types of people.[31] By 1925, according to JoAnne Brown, "the entire public educational system of the United States had been reorganized around the principles of mental measurement," and psychologists had more than seventy-five different intelligence tests, each claiming to measure general mental ability better than the next.[32] For Terman, intelligence was a unified concept, composed of a variety of mental operations that could not be separated out. He noted, for instance, that superior children should be identified through testing because civilization depends on "the advances made by creative thinkers and leaders in science, politics, art, morality, and religion."[33] Later, however, creativity began to be considered separately.

Interest in creativity increased after J. P. Guilford's 1950 address to the American Psychological Association. Guilford argues that creativity had been neglected by psychologists (a state of affairs he found "appalling")[34] and defines creativity as a "relatively infrequent phenomenon" describing individuals with extraordinarily distinguished achievements.[35] Yet, when it comes to actually defining creativity, Guilford cannot avoid circular definitions. For example, he notes that creativity "refers to the abilities that are most characteristic of creative people," that "creative abilities determine whether the individual has the power to exhibit creative behavior to a noteworthy degree."[36] Guilford seems

30. Rose, *Inventing Our Selves,* 57.
31. Terman, *The Measurement of Intelligence,* 11, 13, 17.
32. Brown, 4.
33. Brown, 12.
34. Guilford, "Creativity," 445.
35. Guilford, 444.
36. Guilford, 444.

to have trouble pinning down what he means by creativity and ends up with a circular definition: creative is as creative does.

The definition of creativity most commonly used in psychological and neuroscience research today emerged out of this context. In 1953 Morris I. Stein offered this definition: creative work "is a novel work that is accepted as tenable or useful or satisfying by a group in some point of time."[37] Stein's definition of creativity is a rather rhetorical one. Indeed, Stein stresses the three elements Bitzer would later come to identify as integral to the rhetorical situation: audience, exigence, and constraints.[38]

For Stein, a creative work must satisfy a need for an audience. Stein insists that "the results of the creative process must be communicated to others."[39] Those others form the group that will evaluate whether the creator's work is tenable, or useful, or satisfying or not. This definition is easy to understand when thinking of scientific, technical, or even rhetorical productions. A creative solution to a scientific problem is one that is not only novel but also one that addresses a problem satisfactorily. An audience of fellow scientific researchers, then, is best poised to adjudicate the creativity of a scientific solution or production. A technological work can be assessed by technical experts, but also users. Per Stein's definition, a technological work, such as a new invention, would not be considered truly creative if it did not meet a user's need or gain acceptance by users at some point in time; in that case, it would simply be "novel," but not creative. While the audience for artistic works may be less apparent, according to Stein, in these arts it is the patron or critic, and also the population at large, who determines whether a creative work is tenable or useful or satisfying.[40] For example, art critics and patrons determine what works get displayed in galleries or purchased for home display.

When we say that a work "strikes a chord" with an audience, Stein notes, that is to say that it has fulfilled some kind of need for that audience. For the artist, an audience always "provides the individual with necessary feedback so that he can clarify, alter, or make progress in his future work."[41] Thus, the creative work emerges not out of nowhere but out of a rhetorical situation, including an audience, exigence, and set of constraints. For this reason, studying creative works tell us as much about the audience as the artist, Stein avers, since they can "provide some data for making inferences regarding the needs of the group that are being satisfied."[42]

37. Stein, "Creativity and Culture," 311.
38. Bitzer, "The Rhetorical Situation," 6.
39. Stein, 316.
40. Stein, 320.
41. Stein, 318.
42. Stein, 321.

In Bitzer's definition of the rhetorical situation, exigence is "an imperfection marked by urgency," and it must be one that can be altered through discourse.[43] In his definition of creativity, Stein notes that a creative work arises from "an *interaction* between the creative individual and the problem on which he is working, or, in broader terms, and the environment in which he exists."[44] Stein recognizes that even creative artists (painters, sculptors, dancers, etc.) are working on a problem of some sort. Stein describes the exigence as "a state of disequilibrium" that the creator experiences: "one might say that homeostasis is disturbed, or that there is a lack of closure, or, from a hedonistic point of view, that the individual experiences a lack of satisfaction with the existing state of affairs."[45] Much as the rhetor is compelled to act, in Bitzer's view, by a state of imperfection, so too is the creative person, in Stein's view, compelled by a "state of disequilibrium." In both cases, the creation emerges from a state of affairs external to the creator.

Finally, in Bitzer's definition of the rhetorical situation, the rhetor draws upon "constraints"—a set of factors that both limit and enable a production. Bitzer lists such elements as "persons, events, objects, relations, rules, principles, facts, laws, images, interests, emotions, arguments, and conventions" as among the constraints from which a rhetor can draw.[46] In other words, the rhetor draws on material available to him or her—what Aristotle called the "available means" of persuasion—in order to craft her response. Similarly, Stein notes that creative works do not simply emerge whole cloth; instead, they represent a "*reintegration* of already existing materials or knowledge," even though the resulting creation "contains elements that are new."[47] As in rhetorical theory, Stein's view of creativity stresses the continual reworking of existing materials as a source of novelty. For an artist, these materials might include paint, canvas, found objects, clay, and the like; for a writer they may include extant texts (as in the Shakespeare example above) but also the material tools used for writing and composing (pens, pencils, scratch paper, notebooks, computers, etc.). (In rhetoric, we call this process of working from available materials to create something new *invention,* although our focus on the materiality of that process has not always been foregrounded.)

Thus, Stein presents a view of creativity that is not unlike that which we might expect from a rhetorical theorist. Indeed, his view squares with that of Kenneth Burke, who stressed the rhetorical elements of all arts, includ-

43. Bitzer, "The Rhetorical Situation," 6.
44. Stein, 312.
45. Stein, 312.
46. Bitzer, "Functional Communication," 24.
47. Stein, 311.

ing those deemed "creative" (and hence sometimes situated beyond the purview of rhetoric). In *Counter-Statement*, Burke explains that works of art represent "a particular mode of adjustment to a particular cluster of conditions." Or, in other words, Burke claims that "all competent art is a means of communication."[48] Similarly, in *The Rhetoric of Fiction*, Wayne Booth explains how literary texts can be understood as rhetorical. According to Booth, the writer always guides the attention and interpretations of the reader, despite the common dictum that "true artists" write only for themselves, ignoring the needs of their readers.[49]

Perhaps Burke and Booth would both have embraced Stein's original definition, which suggests that a creative work can only be considered creative insofar as it addresses the needs of an audience. They would also concur that a work of art emerges from with a rhetorical situation or a historical moment, so that it necessarily responds to an exigence.

Thus, in the context of creativity, we might consider whether the creator of a novel work responds to an existing need, or whether he or she helps to constitute that need or exigence, or whether it is a bit of both. From a psychological view, it may be that creativity involves special ingenuity in identifying rhetorical situations not perceptible to others, and devising interesting and effective solutions, or ingenuity in identifying ways to constitute such situations. But Stein's definition of creativity makes creative productions rhetorical. That is, they must address an audience, and do so effectively. The term *appropriate* is sometimes substituted in his clause, and this reflects this rhetorical orientation. After all, appropriateness, or *to prepon*, has historically been the measure of rhetorical effectiveness. The trained rhetor produces not merely novel responses to a situation but appropriate ones, in order to be considered effective.

Crucially, though, the first part of Stein's definition has been widely adopted (and continues to be cited) within neuroscience research, but the last part has dropped out—the emphasis on audience and context, or what rhetoricians might term the *rhetorical situation*. Instead, the definition is most often rendered as "novel and appropriate" or "original and useful," or some variation thereof.

To study creativity, neuroscientists usually rely upon some version of tests that tend to reduce the significance of the rhetorical situation by imposing measures of quantification. Guilford himself developed the Alternative Uses test, which asks people to generate a list of different uses for everyday objects

---

48. Burke, *Counter-Statement*, 107, 73.
49. Booth, *The Rhetoric of Fiction*, 89.

(such as a paperclip or a chair). For instance, if given the prompt of "paper-clip," one might list fishing hook, bookmark, earring, fingernail cleaner, and money clip. The more alternative uses a person can generate, the more creative they are.

In another test, designed by Ellis Paul Torrance, called the "Torrance Test of Creative Thinking," participants are given a battery of different procedures that can be scored according to a coding key. Some of the tests involve verbal responses; for example, in one test subjects are presented with a product (for children, a toy) and asked to identify how it might be improved. A version of the Alternative Uses test is included in the verbal portion. Other parts of the test involve drawing; for example, in one test subjects are given a shape, such as a pear or kidney shape, and asked to complete the rest of the drawing.

By developing these quantifiable tests, Torrance could offer them for use in educative purposes. His goals were certainly lofty; he argued that identifying and nurturing creativity was important to achieve "the American dream of a kind of education that will give every child a chance to grow and to achieve his potentialities."[50] The Torrance test could help educators distinguish the highly creative child from the highly intelligent child. Highly creative children, he argued, were often shunted aside, "considered less desirable as pupils, less well known by the teachers, less ambitious, and less studious and hardworking."[51] Yet these, students, he claimed, possessed the skills that were essential for eminence in almost every field.

Nonetheless, by reducing creativity to a quantifiable trait, the Torrance test (and others like it) constrain creativity—which occurs across various media and fields of inquiry—into a single, measurable entity. In all these cases, the dominant rhetorical operation is synecdoche: to offer a "true definition" of creativity, a part of what we normally mean by creativity in everyday use must be selected to stand in for the whole.

HOW NEUROSCIENTISTS STUDY CREATIVITY

These constrained definitions of creativity materialize in social and cognitive neuroscience studies that use fMRI, in part due to the limitations of the tech-nology itself and in part due to reliance on the definitions themselves. In what follows, I take up two examples of studies that purport to identify neural cor-relates of creativity in two different domains: freestyle rapping and creative

50. Torrance, "Scientific Views of Creativity," 676.
51. Torrance, "Current Research," 312.

writing. These studies can better be understood as studies of improvisation or of what Quintilian termed *facilitas*—"the capacity to produce appropriate and effective language in any situation."[52] Thus, while these studies call what they are studying "creativity," I will use the terms *improvisation* and *invention* here because they provide a more specific description of that which is being studied. Despite their limitations, these studies can be understood in ways that contribute to our understanding of invention as engrained, embrained, and embodied practice.

To operationalize invention, all these studies contrast a test condition involving improvisation with a control task involving copying, or mimesis. The studies hope to identify the neural correlates of creativity by contrasting these two tasks, assuming that the control task does not involve creativity. Yet, I argue here that such an approach overlooks the role of mimesis in training for dance, freestyle rap, and creative writing. Invention emerges not in contradiction to mimesis, but from out of it. Thus, it is a process or practice, not a concrete thing, and should be defined pragmatically. Rather than positing that such studies tell us about creativity, in general, they might tell us about invention as a process embedded in social practices, including artistic traditions and training regimens, embodied habits, rhetorical conventions, and linguistic or corporeal vocabularies and the exclusion of rhetorical situation and exigence.

The Case of the Freestyling Rappers

In 2012 an article appeared in *Scientific Reports* titled "Neural Correlates of Lyrical Improvisation: An fMRI Study of Freestyle Rap." Published by Siyuan Liu et al., the article was a collaboration between researchers at the National Institute on Deafness and Other Communication Disorders, the National Institutes of Health, and Justice by Uniting Creative Energy (JUICE), a Los Angeles–based nonprofit organization that provides training in hip hop arts.[53] Two of JUICE's directors at the time of the study participated in the study development: Michael W. Eagle and Daniel A. Rizik-Baer. According to the report, Eagle and Rizik-Baer were involved in the study design and in behavioral data analysis. Thus, this paper offers an interesting example of collaboration drawing on the expertise of individuals outside the sciences. In addition to being a director of JUICE, Eagle is himself a hip hop performer who uses

52. Fleming, "Rhetoric as a Course of Study," 179.
53. https://juicehiphop.org/

the stage name Open Mike Eagle and has recorded six albums. Rizik-Baer is a music producer and freestyler; he produced the back beat employed in the study and was also a participant.

In the study, experimenters asked twelve male freestyle artists to perform two different tasks in an fMRI. In the control condition, they performed a set of memorized lyrics that they had been given ahead of time, using an 8-bar instrumental track. In the test condition, participants performed a free-style rap over the same 8-bar instrumental track.[54] The experimental setup demands that researchers present a conventional condition that differs from the test condition—they had to oppose something, in this case, to freestyling. The conventional condition has to be "not" the thing being measured in the test condition—thus, "not creativity." Theoretically, one can then simply sub-tract the brain patterns activated in the control condition from those in the test condition to come up with the "neural correlates" of creativity.

The memorized lyrics function as a control that researchers compared to the improvised lyrics. However, anyone who has studied and performed music would say that there is certainly a creative element in performing a set piece of music, as the best concert pianists and singers would demonstrate. Certainly, hip hop artists do not perform the same piece the same way each time, but vary the intonations, cadence, and so on with each iteration. The term "cre-ativity" is reduced to a concrete, specific task, which is necessary to produce evidence within this neuroscientific paradigm, but that reduction glosses over the various ways creativity functions in musical performance.

The scientific article also relies on neurorealism and neuroessentialism. First, the authors must establish that creativity is a "thing" that can be located in or measured in the brain. The title of the article is itself operating on this level. The article's title, "Neural Correlates of Lyrical Improvisation," uses a nominalization, *correlates,* that gives what are essentially correlations a con-crete reality. You can't operationalize creativity empirically in an fMRI scan-ner, seeking how it correlates with activity in different brain regions, without assuming first that it will have a traceable, material reality, as I have shown in chapter 1.

The second rhetorical operation necessary for this type of experiment is choosing a test condition that can represent the mental quality in question. In this case, the authors use "freestyle rap" to represent "creativity." In this very act, then, neuroscientists reduce the whole of "creativity" to a single, concrete example. For this reason, mental concepts are usually defined in sci-entific articles not directly, in categorical definitions, but using synonyms and

54. Liu et al., "Neural Correlates of Lyrical Improvisation," 6.

substitutions, swapping out one term for another. For instance, the authors describe freestyle rap as follows: "Freestyle rap, a popular form, requires an artist to freely improvise rhyming lyrics and novel rhythmic patterns, guided by the instrumental beat—a particularly challenging form of spontaneous artistic creativity."[55] This is a definition by substitution, one that simply swaps one instance, "freely improvising rhyming lyrics and novel rhythmic patterns," with "spontaneous artistic creativity." Creativity has been reduced to "freely improvising rhyming lyrics and novel rhythmic patterns." We are to agree, unequivocally, then, that freestyle rap demonstrates creativity, rather than some other quality, and that it can stand in for other instances of creativity (or at least of spontaneous creativity). We are to accept, further, that the results of this study "may generalize to other forms of spontaneous creative behavior," since, the authors hypothesize, they all involve less activation in the prefrontal cortex, resulting in a pattern where "self-generated action is freed from the conventional constraints of supervisory attention and executive control, facilitating the generation of novel ideas."[56] We must accept the synecdoche, the reduction of creativity to specific neural activations measured by a specific task, in order to accept the experiment's validity.

How might a rhetorical perspective enrich this kind of study and mitigate some of the effects of neurorealism and neuroessentialism? One might point out that a skill such as freestyling is not a pre-existing mental capacity but one that depends on training and socialization. In a *Huffington Post* interview, for instance, Eagle explains that freestyling involves an understanding of meter, of where emphasis and stressed syllables should occur, and of different styles of freestyle. Some, he notes, are more "Dadaist," wherein the rapper just says whatever words come to mind. In others, though, the rapper is always thinking a bar ahead in the music, creating a rough plan on the fly. Often, the freestyle pertains to things that are going on in the moment; thus, Eagle said, he found himself rapping about the fMRI machine and experiment itself during his trial runs.[57] In the context of a freestyle competition, rappers might rap about the situation or about their opponent; "dissing" an opponent, in fact, is a common topic in rap battles, so participants can prepare by generating a repertoire of material they can invoke on the fly.

Further, Eagle noted that he had been freestyling for sixteen or seventeen years, and that there was a definite "mastery arc" involved in learning the craft.[58] The training freestylers describe is, in fact, similar to that required in

---

55. Liu et al., 1.
56. Liu et al., 1.
57. Cronin, "Rappers' Brains."
58. Cronin, "Rappers' Brains."

ancient rhetorical training.[59] For instance, the accomplished hip hop mogul Jay-Z has described his training process in his autobiography, *Decoded*. Jay-Z describes his first encounter with a freestyler, named Slate, whom he found on a street corner "rhyming, throwing out couplet after couplet like he was in a trance." Jay-Z was transfixed by that experience, and he started training himself: "That night I started writing rhymes in my spiral notebook." He would "fill page after page," and he would beat out a rhythm on the table and practice. He'd write everywhere he went, and he says he practiced rap "like it was a sport."[60] For freestylers, writing offers an exercise in fluidity and also provides material for future compositions. In a *60 Minutes* interview with Anderson Cooper, the rapper Eminem displayed one of his boxes of notes—words and rhymes scrawled on yellow legal pads and hotel room notepads—or what he calls "stacking ammo."[61] As rhetoricians, we'd recognize these notebooks as akin to the commonplace books that students of rhetoric have been encouraged to keep in order to have material available for improvising.

Finally, rapping is an embodied activity, but that movement cannot be included when fMRI is being used, since the technology requires participants to keep their heads and bodies rather still. Whereas rap is something that can be done with limited bodily movement, bodily movements play an important role in actual performances. Indeed, in a later interview, Eagle explained that "it was just difficult to keep my body and head perfectly still while attempting to rap. Never knew how much my brain depended on my body to be the metronome."[62] This statement is quite telling, as it suggests that for Eagle, brain and body are intertwined. Thus, what was being measured in the study was not *exactly* the same as the process Eagle uses when he normally raps, as that process depends on bodily movement and rhythm.

Reducing freestyle to creativity also excludes the rhetorical factors involved in such an act. There's a rich body of humanities research into the culture and production of rap that we might turn to. In her study *Hiphop Literacies*, Elaine Richardson describes how hip hop artists develop "an orientation to their situated, public role as performing products" and "that their performances are connected to discourses of authenticity and resistance."[63] The ability to use language creatively, Richardson argues, depends on cultural

---

59. See Hawhee, *Bodily Arts*, for a description of the ancient rhetorical training methods that incorporated gymnastics and wrestling alongside rhetoric.

60. Jay-Z, *Decoded*, 5.

61. Cooper, "60 Minutes."

62. "Speed on the Beat Interviews Open Mike Eagle." http://www.speedonthebeat.com/2016/03/speed-on-beat-interviews-open-mike-eagle.html.

63. Richardson, *Hiphop Literacies*, 1.

rhetorics of survival, discourse patterns with long histories that reflect the itineration of African American people. Successful rappers draw on an extensive vocabulary from standard English, or even other languages, and rhyme them. For instance, in "Rap Protection Prayer," Eagle rhymes "character" with "submariner" and "derringer," and then rhymes "herringbone" with "pheromones" and "baritone." You'll notice here that Eagle has to decide which terms are close enough to a rhyme in order to maintain the cadence of a passage. Some of the best rappers master this skill of choosing almost-rhymes and then delivering them effectively by stressing different syllables, sort of "selling" the rhyme to the audience. (Eminem is famous for this type of rapping, which he demonstrated when he identified five rhymes for "orange" on *60 Minutes*: "orange, four-inch, door hinge in storage" and having "porridge with Geo-rge." He calls this "bending" the word to fit.)

Richardson argues that successful rappers must also use specific terms that commonly appear in freestyle, each of which has a unique derivation and history of use. For instance, the term *flow* refers to "the way that a rapper solidifies his/her vocals rhythmically and sonically within the track or the beat," not just to the more popular understanding of flow as being in a creative zone.[64] To produce freestyle rap that will be legible with an audience of freestyle aficionados, rappers need to draw on an extensive vocabulary that includes these specific terms. These might include novel coinages, such as "crunked up" (used to describe a high-energy or party atmosphere); new uses of standard English terms, such as "percolate" (a dance), "hood" (neighborhood), "crib" (one's home); or modifications of existing terms (such as "diss," short for "disrespect").[65] The rappers in the scientific study drew on a rich vocabulary from years of honing their craft. That training makes improvisation possible.

The Case of the Creative Writers

The second case to consider here is a 2013 study called "Neural Correlates of Creative Writing: An fMRI Study," conducted by Carolin Shah et al. This study differs from the previous one in that it did not use experienced artists, but instead used twenty-eight individuals with no writing experience to do creative writing in an fMRI.[66] In the study, the participants were asked to complete three tasks: the first was to copy part of a given text for sixty sec-

64. Richardson, 107.
65. Richardson, 106–7.
66. Shah et al., "Neural Correlates of Creative Writing: An fMRI Study," 1090.

onds; the second was to brainstorm ideas about a passage, "generating ideas for their highly creative continuation"; and the third was to actually write a "new, creative continuation of the same given text."[67] Thus, the test sought to distinguish between three different modes of activity using a technique similar to that used for the freestyle rappers. However, they also included a "brainstorming" mode, one not granted to the rappers.

This study replicates the rhetorical structures of the other study, however, insofar as it relies on a fundamental contrast between an activity not deemed creative, in this case copying a text, and one deemed creative, in this case, continuing a text by actually writing it in a "creative" fashion. Here again, one task is taken to stand in for "creativity" through the operations of synecdoche. This is the case even though the authors state that they "doubt that the large spectrum of creative writing properties depends on a single mental ability."[68] Thus, even though they recognize creative writing as a multifaceted activity in principle, they reconstitute it as a single thing in practice, both in the experimental design and in the overall rhetorical structure of the article.

As was the case in the rap study, a copying task was given first, as a kind of control. Yet, in creative writing as in rap, copying serves an important part of the pedagogical process. For instance, one might be asked to rewrite a text from the eighteenth century using contemporary language, or to compose a poem using only words and phrases found in another text or set of texts, or to replicate the form of a story (phrasing, perspective, and structure) while writing about an entirely different topic or theme. (These kinds of activities will also be familiar to rhetoricians as a form of the ancient art of *imitatio.*) All these mimetic activities are actually part of learning to write creatively, and are therefore not "uncreative" in and of themselves.

Further, as Paul Prior and Jody Shipka have demonstrated, writers draw upon complex networks of material and temporal practices, or "chronotopic laminations," to manage consciousness and enable what neuroscientists might call "creativity." These practices include the physical locations where writers work (which may vary from a home office to a coffee shop to a library or dorm room), the bodily positions writers prefer (standing, or sitting, or reclining in bed, for instance), the materials they prefer (which may include writing tools such as a specific type of pen, computer software and hardware, and even "peripherals" such as a cup of coffee), and temporal factors (such as setting a timer so that the writer composes for a particular amount of time before get-

---

67. Shah et al., 1090.
68. Shah et al., 1098.

ting a break).[69] Those material and temporal elements fall out of the picture when subjects are writing while lying down in a sterile, uncomfortable scanner. (Note, for instance, how the "conventional felt-tipped pen" is presented as a default in the experiment but would be constituted as a specific material practice in Prior and Shipka's study.)

A trained creative writer, like the trained rapper, has a repertoire from which to draw. When asked to continue a story from a given passage, for instance, they might draw on archetypal plots (such as the journey or quest motif) and characters (the hero, the villain), or they might choose to flout those archetypes. Experienced creative writers have some knowledge of plot structures, setting, character, genre, and dialogue, all of which provide patterns that they could either draw from consciously or flout. But even inexperienced writers likely have some implicit knowledge of these concepts, either from study of literature in English classes, or from their own exposure to narratives in books, television, and film.

In fact, in the supplementary materials provided with the Shah et al. article, the authors provide two examples of texts composed by participants. In the first, the writer clearly draws on the genre of comedy, describing how a man's violin was eaten by a donkey so he decided to play the comb instead. In the second, the writer draws on the genre of crime or horror, creating a more macabre narrative about a murder-suicide involving a man entering a Laundromat and delivering items covered with his wife's blood, and then killing himself by sticking "his head between the dryer coils."[70] To compose such a response, the participants might have consciously activated that genre knowledge, or they may have done so unconsciously; from the study design it is impossible to know which.

By using only untrained writers, the study by Shah et al. overlooks the complex factor of training, craft, and expertise. Thus, it might be better described as a study of the "neural correlates of creative writing in novices." Yet, the authors address this limitation by stating that "writing texts or stories is a common human ability," implying that there must be some underlying similarities between novices and experts. Again, creativity is presented synecdochally, as something that can be measured and defined through a single, concrete task, and isolated from other processes.

---

69. Prior and Shipka, "Chrontopic Laminations."

70. Shah et al. Available as a link from the online article at wiley.com.

A NEURORHETORICAL APPROACH TO CREATIVITY

To address essentialism, interdisciplinary or transdisciplinary research teams can employ the following heuristic. This heuristic is intended here as a way of opening possibilities that may not have been considered previously. It may also be used by individuals seeking to interpret neuroscience studies for their own work in other disciplines.

Involve Experts

First, teams might ask: who else can we involve? For scientists, it is common to assemble an interdisciplinary team to work on a common task, but it is less common to think about humanities scholars as part of the team. In the cases described above, a scholar of hip hop or creative writing respectively might have been included on the research team. I might argue that rhetoricians, as well, can contribute usefully by asking broader heuristic questions from the perspective of someone who studies disciplinary communication.

Clarify Definitions and Their Implications

As a next step, research teams can clarify definitions of the concept under study by highlighting how a given topic might be understood within their discipline. For instance, while the studies above both rely on the topic of "creativity," in rhetoric we might consider these in relation to the canons of invention, arrangement, style, memory, and delivery—all of which were trained capacities that led to a rhetor's ability to speak extemporaneously or to perform a speech effectively. The term *creativity*, a rhetorician might point out, does not capture the idea that improvisation is a trained capacity. Using the term "invention" instead of creativity might lead to different kinds of studies in which participants were engaging with source material, brainstorming, and so on as part of "creativity" itself, rather than its opposite.

For the freestyle rappers (and to some extent the untrained creative writers), improvising test subjects were drawing on knowledge that had likely become habituated. This leads to one change in interpretation: the patterns of brain activation may indicate not "creativity" per se, then, but a mode of action involving habituated responses and trained capacities that do not require "executive control" because they are deeply engrained. In rhetorical terms, we might describe these as studies of *facilitas,* not "creativity" writ large.

But more broadly, this perspective calls into question the methodological separation between rote repetition and improvisation (or creativity) that each study assumes. In rhetorical training, repetition and imitation are important exercises in developing the ability to improvise. One exercise, a version of paraphrase called pattern practice, asks students to analyze the structure of a given sentence and then to write their own sentence using the original one as a model, copying the same grammatical and rhetorical pattern.[71] In Renaissance times, these exercises would have been done also by translating sentences from one language to another, say, from Latin to English or from Greek to Latin to English.[72] Students would also keep commonplace books, where they copied quotations from other works to use later as inspiration for their own compositions. All these exercises gave students a repertoire of linguistic and rhetorical strategies from which to draw creatively. As W. Ross Winterowd writes, "Stylistic exercises enable. That is, 'mere' exercises in style allow the student to internalize structures that make his own grammar a more flexible instrument for combining and hence enable the student to take experience apart and put it together again in new ways, which is, after all, the generative function of language."[73] "Mere repetition" and copying, then, lead to novel products—the very definition of creativity that neuroscientists seek to uncover.

Debra Hawhee explains how the ancient training program in rhetoric drew inspiration directly from the *gymnasmata,* or program of physical exercises—and that, in fact, the two types of training sometimes occurred concurrently, in the gymnasium.[74] Hawhee asserts that this common location led to a "crossover in pedagogical practices and learning styles" in rhetoric and in athletics,[75] so that rhetorical training took on the trappings of athletic training; namely, a pedagogical model that emphasized "rhythm, repetition, and response."[76] Like the rhetors Hawhee describes, creative artists have hours of repetition under their belts. Rappers might repeat their own lyrics, or those of another artist, again and again to develop skills in delivery—a skill brilliantly displayed in a viral video that features prominent rap artists freestyling based on the words to the children's book *Llama Llama Red Pajama.* Although given the same words, each rapper uses different strategies to dramatize the book. Ludacris uses intonation, rhetorical questions (rendering "Llama Llama Red

---

71. Corbett, "The Theory and Practice of Imitation in Classical Rhetoric." 248.

72. Corbett, 247.

73. Winterowd, "Style: A Matter of Manner," 164.

74. Hawhee, 110.

75. Hawhee, 111.

76. Hawhee, 135.

Pajama" as "Llama Llama—what?—Red Pajama," for instance), reduplication (changing "Mama says she'll be up soon" to "Mama says she be up soon, be up soon soon"), and gestures to render the children's story in hip hop style.[77] In his version, Lil Yachty provides an opening and closing dedication, in the style of hip hop recordings, ending with a shout out to "all the baby mamas" and to "all the young kids."[78] In each case, the rapper in question draws on rhetorical conventions and remixes them to create a unique, "creative" version of the same text. Yet in the studies I have reviewed here, imitation and invention are opposed, with the production of a rote exercise being used as the counter to the improvised performance.

The goal of comparing definitions should not be to arrive at one "true" definition of a given concept but to reflect on how those concepts are in fact enacted in particular settings, or to acknowledge the inevitable multiplicity of concepts like creativity. Rather than choosing a universal definition, it might be appropriate to acknowledge the range of definitions available and to specify why a given definition has been chosen as well as the exclusions that definition entails.

Reconsider Study Design

Given these alternative definitions, research teams might develop study designs that could provide a finer-grained or simply alternative hypothesis about the concept in question. For instance, by drawing on the perspectives of someone trained in rhetoric and/or hip hop, the researchers in the first example may have calibrated their study of creativity and what it purports to measure. They might have specified that they were studying improvisation, not creativity writ large, and they might have tailored the tasks to provide a more fine-grained analysis. For instance, the experimental apparatus sets up the two tasks as opposites—one involving improvisation, one involving rote repetition. But in between those two tasks, one might place a task asking freestylers to *creatively* perform a simple rap they had memorized, a task that would require participants to imagine themselves varying the rhythm, phrasing, performance quality, or feel of the words. Or, they might ask the freestylers to imagine different audiences or situations in different test conditions. Similarly, the creative writers might be asked to do an imitation study

---

77. Ludacris, "Llama Llama Red Pajama Freestyle."
78. Lil Yachty, "Llama Llama Red Pajama over Ms. Jackson Beat."

that involved repurposing existing language, such as copying the structure of a poem with new words and ideas.

Resituate the Study

Finally, when presenting the results of these kinds of studies, researchers might draw more deeply on humanistic theories in the introduction and discussion sections of a report. While an extended analysis is not often possible within the limited page counts for journal articles, it is still possible to reference some sources that contextualize or broaden the discussion. In the freestyle-rap study, none of the citations in the reference list refer to scholarship about hip hop itself. Doing so might lend some texture and context to the claims the article makes about the practice, as I have suggested above. In the creative-writing study, two out of forty-nine citations refer to writing—including one from the field of rhetoric and composition, a reference to a now-canonical 1981 article by Linda Flowers and John Hayes on a cognitive process theory of writing.[79] While important to the field of rhetoric and composition, however, this article reflects an early stage in the attempt to understand writing as a process and has been extended significantly by work that seeks to contextualize that process in terms of materiality, space, and time;[80] enrich the field's understanding of invention;[81] and consider the role of technology in shaping writing practices,[82] to name just a few relevant considerations. These considerations could lead to further research questions and could also indicate some of the limitations of studies that must necessarily isolate individual writers in individual fMRI machines rather than allowing them to move around, interact with others, and so forth. Importantly, fMRI studies of improvisation or composing enact a model of human agency that is self-contained, where composers are autonomous agents unaffected by their surroundings, materials, or other people. Current scholarship in rhetoric and composition would call this into question, suggesting a more distributed model in which agency tracks across the experimental apparatus, the physical and temporal context of the study, the materials provided to participants (from a rap track to a felt-tipped pen), to the researchers themselves and their position and role in relation to the participant. It would be interesting to consider what such a view of composition might entail for a research method such as fMRI that is predicated

---

79. Flower and Hayes, "A Cognitive Process Theory of Writing."
80. Prior and Shipka, "Chronotopic Lamination."
81. Muckelbauer, *The Future of Invention*; LeFevre, *Invention as a Social Act.*
82. Moore et al., "Revisualizing Composition."

on the separability and agency of the human brain. Aside from multiplying variables—type of rap track, felt-tip pen or pencil, the instructions given to participants and by whom, and so on—it would be interesting to consider how to integrate multiple studies of composition and improvisation to generate a more holistic understanding of brains as part of the enacted material-discursive entanglements at stake in the experiment.

In this chapter my main goal has been to show how neuroessentialism arises from experimental settings, not just from popularizations of social and cognitive neuroscience, and to suggest a heuristic that can be used both to interpret neuroscience studies and to identify alternative paths forward. In the next chapter, I'll show how this approach can be combined with the heuristic developed in chapter 1 to address neurorealism as well as neuroessentialism. This approach, I argue, could enhance neuroscience studies of persuasion—a phenomenon of interest to rhetoricians who may just have a few ideas about that topic as well.

CHAPTER 3

═══════════

Neurorhetoric

"THANKS TO advanced brain scanning techniques, we've learned more about persuasion than we've ever known"—at least, that's what Carmine Gallo, author an article titled "Neuroscientists Say This Tool Is the Secret to Persuasion," would have us think. Gallo, who purports to have studied persuasion for over twenty-five years, claims that, thanks to brain science, we can now pin down "what works and what doesn't" when attempting to persuade. What works, Gallo writes, is "telling a story," a tactic that activates the amygdala, the brain's "ticket to persuasion." By now, it should be easy to recognize both neurorealism and neuroessentialism at work in this example. First, Gallo offers a neurorealist metaphor, positioning a region of the brain as a concrete object (or "ticket") to persuasion. Digging deeper, we see that, in doing so, Gallo must essentialize persuasion by reducing it to a particular thing. As Gallo puts it, "Telling an emotional story is the single best tool we have to make an instant and powerful connection with another person."[1] Here, then, persuasion is reduced to an emotional connection with another person, or what Kenneth Burke would refer to as identification.

While rhetoric scholars are likely to view persuasion as their unique domain, as this example shows, the topic has been of interest to scholars in a range of fields, including psychology, law, consumer marketing, and neurosci-

---

1. Gallo, "Neuroscientists Say This Tool Is the Secret to Persuasion."

ence. While they share an interest in questions of the influence or effectiveness of language, each discipline goes about studying persuasion differently. Given these differences, and the risks of neurorealism and neuroessentialism, what would a productive way of examining persuasion look like?

Here, I'll focus on how neuroscientists and rhetoric scholars study persuasion, identifying some similarities and differences between the two. In fMRI studies, neuroscientists typically treat persuasion as a discrete event in which an individual is "exposed" to an argument and then is either convinced or not. Similar assumptions guide scientific studies of the persuasiveness of neuroscience itself, of which there have been quite a few. Instead, I suggest, rhetorical theories would push neuroscientists to understand persuasion as occurring over time, through repeated, serialized encounters with arguments, tropes, and affects that become embodied through a process of circulation and accretion.

In this chapter, I'll first examine several neuroscience studies that purport to identify neural correlates of persuasion. In these studies, researchers assume that an effective argument shares specific features that will persuade audiences immediately and universally. This assumption shapes the materials and methods of the experiment and, therefore, the results, offering a narrow, bounded view of persuasion. In effect, these studies rely on neuroessentialism and neurorealism in ways similar to that of other examples we have examined thus far. This same view of persuasion, I'll show in the second section, shapes neuroscientists' attempts to understand the persuasiveness of their own field of study—of the neuroscientific images and language that pervade public discourse. Finally, I'll suggest a method by which one might investigate persuasion by raveling together humanistic and neuroscientific perspectives to develop a rich area of mutual inquiry into the phenomenon of persuasion.

HOW RHETORICIANS AND NEUROSCIENTISTS STUDY PERSUASION

The study of persuasion in rhetoric and the study of persuasion in psychology were not always separate. Those interested in rhetoric have always been (by necessity) interested in the mind. Indeed, Aristotle's analysis of rhetorical appeals offers a prototype of faculty psychology, an area of study carried out by nineteenth- and early-twentieth-century rhetoricians such as George Campbell, Gertrude Buck, and Hugh Blair. Later in the twentieth century, though, each discipline narrowed its focus, beginning to develop somewhat separately, even if those on the rhetoric side of things continued to borrow from psychological theories such as those of William James, Sigmund Freud, Edward B. Titchener, and Carl Rogers, and even if psychologists continue to cite Aristotle and Plato, at least superficially.

Both fields have grappled with similar questions but in different ways. How do we account for *ethos,* or the credibility of the speaker? How do we account for the persuasiveness of arguments on different topics? And how do people come to believe what they believe? While these questions may be similar, the methods used by both disciplines differ. Rhetoricians tend to delve deeply into an event or case, closely examining a persuasive speech, or an event, or a debate and the context surrounding it, using careful analysis of texts, images, and other artifacts (increasingly, sound, video, and objects). We have also tended to be promiscuous in our theoretical engagements, drawing from an ever-widening engagement with philosophy, feminist studies, critical theory, disability studies, critical race studies, new materialisms, field methods, and so on. While rhetoricians still engage with the question of how and why something is persuasive, our answers become increasingly layered, complex, and tentative as we recognize that, often, no single, universalizing answer can be given.

Psychologists, meanwhile, take on the apparatus of the experiment, creating controlled situations in which individuals can be exposed to different kinds of arguments and their responses monitored via questionnaires or, increasingly, brain imaging technologies. Thus, while rhetoricians value textual detail, complexity, and ambiguity, psychologists tend to value generalizable types of arguments, clarity, and certainty. One indication of this may be that psychological studies often omit the very persuasive texts used as materials in the study, or, at most, provide one or two examples. A rhetorician would surely want to get their hands on the specific texts used as prompts in a study.

One psychological view of persuasion used in social and cognitive neuroscience studies (especially those that use fMRI) is simply this: strong arguments are persuasive. A strong argument is logical, appealing to one's sense of reason; emotionally simulating, appealing to a person's sense of emotion; and credible, appealing in part through the use of an authoritative speaker or writer. (It is not difficult to see the connection to Aristotle's pisteis here: *logos, pathos,* and *ethos.*) Weak arguments are those that do not do these things. Presumably, a rational person should be persuaded by a strong argument. From this view of persuasion, it would follow that one can study persuasion in the brain by exposing people to "strong" or persuasive arguments while in an fMRI machine.

The limits of this view, though, become apparent as soon as you step out of the lab and into the world outside of it. If "strong arguments" are persuasive, why is it that only 50 percent of Americans believe that climate change is happening because of human activity, despite the preponderance of scientific evidence supporting that argument?[2] Why do 9 percent of Americans continue to doubt that childhood vaccines are safe, despite the overwhelming evidence disproving a link between vaccines and autism?[3] And why do 31 percent of Americans believe that humans have not evolved over time but that they "have existed in their present form since the beginning of time"—despite the bountiful scientific evidence supporting evolution?[4] Views of persuasion grounded in the theory of "strong arguments" simply fail to explain why so many are not persuaded by them, even though these theories seem to allow for factors such as credibility, emotion, and expertise.

Survey data suggests that what you think about climate change, or evolution, or childhood vaccines has much to do with your political affiliation, age, gender, education, religion, and/or race and ethnicity. For instance, Americans over the age of sixty-five are more likely to doubt human influence on global warming, as are those without a college degree and those who align themselves with the Republican Party.[5] When it comes to vaccines, however, older Americans are more likely to regard them as safe, and moderates are more likely than either conservatives or liberals to agree.[6] Gender comes into play in views of human evolution, with women being more likely than men to believe either that humans have not evolved over time or that evolution was guided by a supreme being.[7] This survey data shows, then, that "strong arguments" do not automatically persuade everyone. We can interpret this data to suggest that reason, emotion, and credibility are not universal appeals but depend on the audience—an insight so obvious as to nearly go without saying. But beyond that, we might also say that persuasion extends beyond a single, bounded "strong argument" and is diffused throughout the social and material worlds we inhabit. Social locations (gender, class, race, political orientation, and so on) affect the kinds of arguments and topoi to which one is exposed. Physical locations—such as living in a suburb versus a city—may also affect opinions. We might add a host of rhetorical factors to those outlined in the traditional tripartite scheme of *ethos, pathos,* and *logos,* such as sensation, affect, embodiment, time and space, and the circulation and sedimentation of rhetorics in networks or clouds. Thus, while at one time rhetoric scholars may have agreed with the psychological approach to persuasion—at least in general

2. Funk and Rainie, "Americans, Politics and Science Issues," 37.
3. Pew Research Center, "83% Say Measles Vaccine Is Safe for Healthy Children."
4. Funk and Rainie, "Americans, Politics and Science Issues," 87.
5. Funk and Rainie, 39–40.
6. Funk and Rainie, 121–22.
7. Funk and Rainie, 90.

terms—rhetoric has always been invested in contingency, or the understanding that situations are important. Today, our understanding of the various factors making up a rhetorical situation has expanded considerably.

To start, though, I wish to focus on the materiality of fMRI-based studies of persuasion: the elements used to produce "persuasion" in an fMRI and then to measure its neurological effects. The problem with a narrow view of persuasion is that it is based on a fabricated understanding of "strong arguments"—that is, that experiments based on this approach rely on artificial materials and limited assumptions about rhetoric. Overall, the methods used for fMRI studies of persuasion display the following psychological techniques and assumptions: 1) that persuasion can be neatly defined and delimited, 2) that persuasion can be "tested" by offering participants discrete exposures or prompts in an fMRI machine, and 3) that persuasion can be identified in a discrete moment when a participant concurs with an argument.

First, researchers tend to define persuasion as though it were a fairly simple and obvious phenomenon. For example, in a 2010 study titled "Predicting Persuasion-Induced Behavior Change from the Brain," Emily B. Falk and her colleagues define persuasion by example:

> People are exposed to an almost endless stream of persuasive messages each day: television and magazines are full of explicit advertisements, friends and family induce us to see things their way, and even educators fill their teachings with rhetorical devices meant to persuade.[8]

These instances are presumed to be relatively straightforward examples of persuasion, which itself requires no further definition.

In other studies, researchers draw on previous findings and psychological theories of persuasion as a sort of definition. For instance, in their 2013 study, "Affective and Executive Network Processing Associated with Persuasive Antidrug Messages," Ian S. Ramsay et al. do not explicitly define persuasion, but they use previous research to set out an implicit definition. First, Ramsay et al. note that persuasiveness depends on "emotional arousal," citing previous research, including that of Falk and her colleagues. Then, they pose the question of whether persuasiveness also involves "executive brain areas" (a shorthand for reason), since the existing research has shown that "too much affective engagement during persuasion" can overwhelm the brain, making an

---

8. Falk et al., "Predicting Persuasion-Induced Behavior Change from the Brain," 8421.

argument less effective. Thus, they argue, "successful persuasion attempts will involve more than just socioemotional systems"—namely, "executive control."[9]

To support this claim, they cite a model of persuasion developed by Richard E. Petty and John T. Cacioppo, the "elaboration likelihood theory of persuasion," which actually suggests two kinds of persuasion. The first is based on careful though and reflection, or elaboration; the second occurs when people are swayed by other cues (emotions, expertise, etc.) and do not engage in careful scrutiny of an argument.[10] It is pretty clear from the way these are described that the former type of persuasion is considered better than the latter: when conditions encourage elaboration, they write, people "attend to the appeal; attempt to access relevant associations, images, and experiences from memory; scrutinize and elaborate upon the externally provided message arguments . . . and draw inferences" and ultimately arrive at a well-founded attitude toward the message.[11] This model also seems to assume that the rational, logical elements of persuasion can somehow be uncoupled from the nonrational emotional or ethos-driven cues, a view that rhetoricians would contest.

By drawing on previous research in the introduction to the article, then, Ramsay et al. develop a definition of persuasion in which emotional arousal is necessary but not sufficient; some kind of evaluative reasoning must also occur. Evidently, this definition serves their rhetorical purposes in the introduction to the experiment, which is to establish an exigence for their study of the extent to which "executive control" and affective factors influence participants' responses to arguments.

Both Falk and Petty were interested primarily in what rhetoricians would term the *logos* and *pathos* appeal in arguments. Other studies have taken up what we would term *ethos* by studying how participants respond to photos of celebrities that are linked to different kinds of products (in one case, shoes). In a 2008 study, Vasily Klucharev, Ale Smidts, and Guillén Fernández sought to replicate the kind of persuasion that occurs when people view print or television ads featuring a celebrity endorsing a product. Klucharev, Smidts, and Fernández define persuasion in equally straightforward terms as "any message that is intended to shape, reinforce, or change the responses of others."[12] Persuasion is unidirectional in this view; a speaker or writer creates a message to affect others.

---

9. Ramsay et al., "Affective and Executive Network Processing Associated with Persuasive Antidrug Messages," 1137.

10. Petty and Cacioppo, "The Elaboration Likelihood Model of Persuasion," 125.

11. Petty and Cacioppo, 128.

12. Klucharev, Smidts, and Fernández, "Brain Mechanisms of Persuasion," 353.

Persuasion works in ads, they argue, but not primarily through *logos*; that is, ads seldom offer an "explicit message" but instead simply show a celebrity alongside the object in question.[13] These kinds of ads instead invoke the expertise of the celebrity; the authors give the example of Bill Cosby, who until recently was considered a trustworthy father figure and therefore an appropriate spokesperson for Jell-O.[14] Expertise, in this view, is understood as something a person either does or does not possess in relation to a product; it is largely a property of a person rather than an emergent quality that depends on the rhetorical situation and a network of effects and affects. During the time Cosby was considered an exemplar of *ethos,* he was in fact drugging and assaulting women—he was not in fact an ethical person. Yet, Cosby only came to be lacking in *ethos* when those assaults were made public much more recently, which was enabled in part by media networks including online gossip websites, tabloids, and mainstream newspapers. *Ethos,* then, is rarely as straightforward as is suggested in Klucharev, Smidts, and Fernández's study of celebrity endorsements.

Typically, researchers take persuasion to be exemplified in a series of short texts that make an argument. They ask participants to read (and sometimes respond to) a set of such arguments that have been generated by researchers; sometimes, these arguments have previously been rated as strong or weak by an earlier set of participants. Then, they record fMRI data as participants read the arguments. Next, the fMRI data may simply be analyzed using statistical methods to determine which brain regions were activated, or it may be correlated with the participants' later ranking of how persuasive they found those statements, or it may be correlated with later behaviors. Regardless, this study design assumes that persuasion can be identified in the single moment when participants were "exposed" to the argument and the fMRI scan occurred. Researchers must determine exactly how long to give participants to absorb the argument, generally on the order of seconds.

For instance, Falk et al. employed slides that contained information about the importance of sunscreen use that was drawn from "information from reputable organizations."[15] Each slide provided an argument for sunscreen use, such as this one given on a sample slide:

UV rays reach you on cloudy and hazy days, as well as bright and sunny days. UV rays will also reflect off any surface like water, cement, sand, and snow. Therefore, it is always important to wear sunscreen. Some people

---

13. Klucharev, Smidts, and Fernández, 355.
14. Klucharev, Smidts, and Fernández, 354.
15. Falk et al., 8422.

think that it is only important to wear sunscreen on days when they go to the beach, but to maintain healthy, happy skin long into old age, it is important to apply sunscreen on any day when you will be outside. Wearing sunscreen consistently is the best way to promote attractive and healthy skin!

Participants had between 15 and 37 seconds to listen to each message being read to them as they followed along with the text portrayed on screen. They heard and read six messages while their brains are being scanned. Thus, they were likely exposed to such messages for about 6 minutes total, at most. With this study design, Falk et al. purport to be able to recognize that "neural responses can be recorded *in the moment persuasion occurs.*"[16] In this way, the Falk study assumes that persuasion happens relatively suddenly, in a particular moment, not as an accretion of beliefs that shift over time. Such an approach also assumes that participants are a sort of tabula rasa, with no previous exposure to information about a topic.

While rhetoric scholars recognize the importance of timing, or *kairos,* scholars also recognize that persuasion occurs not simply in a single moment but often over the course of a lifetime. Burke writes that "often we must think of rhetoric not in terms of some one particular address but as a general body of identifications that owe their convincingness much more to trivial repetition and dull daily reinforcement than to exceptional rhetorical skill."[17] By privileging single, discrete arguments presented at a particular moment, Falk's fMRI study does not account for this body of identifications.

In the example of sunscreen use, it may not be any one advertisement or public service announcement but the broader, repeated exposure to such ideas that convinces someone to wear sunscreen. For instance, we may see advertisements for sunscreen on television, encounter sunscreen on the shelves while shopping at the pharmacy, see others applying it at the beach or pool, read articles about skin cancer in magazines, see social media posts about which sunscreens are safest for children, or recognize the unmistakable scent of coconut and chemicals at the pool and be reminded to reapply. We may apply sunscreen to our bodies on a sunny day and notice that we don't get burned. And while, in theory, everyone should wear sunscreen, we see those products being targeted especially to white people and particularly to women and children. Some individuals are more likely to be interpellated by sunscreen arguments than others. Together, this set of identifications contributes

---

16. Falk et al., 8421; emphasis added.
17. Burke, *Rhetoric*, 26.

to the message that it is important to wear sunscreen (for some people)—probably more so than a single exposure to a single message.

The assumption that persuasion happens in a single instant, however, suits the experimental apparatus used for fMRI studies. The "body of identifications" that Burke describes would be difficult to measure in such a setting—there are many different messages and instances to choose from. One could not reasonably ask participants to recall every message they had ever seen about sunscreen use and then to rank them individually while lying in a scanner. Nor could you have someone remain in a scanner for hours while you showed them every possible type and form of sunscreen argument, from the most explicit to the most mundane, spraying Coppertone in the lab or showing advertisements or listening to snippets of conversation about skin cancer recorded in a coffee shop. That would not be an easy (or accurate) undertaking. Thus, the experimental apparatus for fMRI experiments itself dictates how we have to understand persuasion; in this case, as something that happens in the discrete time-space of exposure to an ad or message.

To study persuasion, researchers also typically rely on various forms of quantification. For instance, in Ian S. Ramsay et al., participants watched a 30-second antidrug clip from a public service announcement. These clips had already been rated as either strongly or weakly convincing by independent viewers. (Presumably, those that were only considered moderately convincing were excluded.) The clips chosen included features such as "negative consequence frames, intense imagery, narrative structure, social appeals, and plot twists or surprise endings."[18] Study participants then watched the videos again and rated them on a 7-point scale from 0 (*bored*) to 7 (*stirred up*); they also rated them on a 9-pronged scale of convincingness:

- Extremely unconvincing to extremely convincing
- Extremely unbelievable to extremely believable
- Extremely forgettable to extremely memorable
- Extremely bad to extremely good
- Extremely unpleasant to extremely pleasant
- Extremely negative to extremely positive
- Extremely not for someone like me to extremely for someone like me

Together, these ratings were "summed to form a measure of perceived convincingness."[19] In this study, we see a discourse of quantification at play, an

---

18. Ramsay et al., 1138.
19. Ramsay et al., 1138.

assumption that a clip is either persuasive or not persuasive, regardless of the audience and context.

Quantification, as Jessica Mudry has shown, tends to strip that which is being quantified from its qualitative elements, rendering unique entities interchangeable.[20] When persuasive messages are quantified, we lose what is particular about each message—its particular use of language, tropes, topoi, emotional appeal, images, and so forth—in favor of an independent number or loose set of features such as "intense imagery" or "social appeals." Of course, neuroscientists are less interested in these particularities than they are in uniformity. As a humanist, though, I can't help but wonder what may be missed when persuasion is understood in this narrow, quantified way. For example, are "moderately persuasive" messages potentially more persuasive in the long term than those that are immediately persuasive? Can a persuasive message mix some of the qualities listed above? Can they be unpleasant yet persuasive in some way—if not rationally than emotionally?

In the studies I have examined so far, the arguments used were at least based on ones that the authors had found in other places—real-world arguments. In other cases, though, the arguments used are carefully selected, sanitized, rated, quantified—not real-world arguments. For instance, to develop the Elaboration Likelihood model of persuasion, Richard Petty and J. T. Cacioppo began by "generating a large number of arguments, both intuitively compelling and specious ones, in favor of some issue." Next, they asked "members of the appropriate subject population" to rate those arguments on a scale of persuasiveness. Based on those ratings, they selected the arguments with the highest and lowest ratings. Finally, they asked another group of participants to record their thoughts related to those messages, which were then coded as favorable, negative, or neutral. A "strong message," they conclude, is one that is highly persuasive *and* one that produces favorable thoughts.[21] While not all fMRI studies use this particular model to generate arguments, they tend to share similar assumptions about persuasion; namely, that messages can be created by researchers, ranked and evaluated, and then deemed uniformly "strong" or "weak," "positive" or "negative"—despite the fact that in practice, we know people respond in dramatically different ways to the same arguments.

In some cases, researchers seeking consistency schematize arguments to the point that the prompts barely resemble a persuasive appeal that would be found outside of the laboratory. Klucharev and Smidts sought to emulate the

---

20. Mudry, *Measured Meals*, 174.
21. Petty and Cacioppo, 133.

effect of a celebrity endorsement that would appear in a magazine or television ad. To do so, they "presented young female subjects interested in celebrities and shopping with photographs of famous persons followed by photographs of everyday objects."[22] The researchers claimed that their study design "imitates commercials where celebrities present (communicate) certain products or information" with very little explicit argument or connection supporting why that celebrity endorses that product (or why that celebrity has any special credibility for that product type).[23] While it is true that such ads exist, they do not generally feature just a picture of a celebrity and a picture of an object on a blank page.

In their analysis of celebrity images in magazine advertisements, Marla Royne Stafford, Nancy E. Spears, and Chung-Kue Hsu indicate that this subgenre of ad adheres to Gestalt principles of cohesiveness. That is, designers create visual and verbal links between images of the celebrity and the product so that a connection can be made between the two on the basis of "similarity, interaction, proximity, familiarity," and so on.[24] In doing so, designers try to communicate a message to the audience—it is not simply the case that viewers of those ads have to construct the connection between the two out of thin air.

Take, for instance, an ad for Chanel No. 5 perfume featuring a black-and-white image of Brad Pitt (with goatee and long hair) and a color image of the bottle (Figure 1). Chanel No. 5 is a perfume for women. So why does it feature a brooding Brad Pitt staring intensely into the middle distance? Why does it feature simply the word *inevitable*? The connection here is not just visual but rhetorical and cultural. Viewers may be expected to (consciously or not) fill in what they know about Brad Pitt—that he is known as a handsome movie star who met his then current wife, Angelina Jolie, on the set of a film, where their chemistry was, perhaps, "inevitable" (leading Pitt to divorce his first wife, Jennifer Aniston). Is Pitt's smoldering stare in this image meant to evoke that inevitable chemistry with Jolie (and, perhaps by implication, the smoldering stares one might also expect while wearing that perfume)? Notice also that following Brad's line of sight returns the reader's gaze to the bottle of perfume. The design of the ad, then, creates a loop, where the reader's gaze shifts from the bottle of perfume, to Brad, and then back again.

The Gestalt view offers one interpretation of how advertisements persuade, but of course they do not persuade uniformly. In practice, ads are planned,

22. Klucharev, Smidts, and Fernández, 354. The assumptions in this phrase about sex and gender are also noteworthy and might be considered in relation to the material in chapter 4, "Neurosex."

23. Klucharev, Smidts, and Fernández, 354.

24. Stafford, Spears, and Hsu, "Celebrity Images in Magazine Advertisements," 15.

FIGURE 1. Perfume ad for Chanel No. 5 featuring Brad Pitt (Source: https://
therefinededit.files.wordpress.com/2012/10/brad-pitt-chanel-no5-01.jpg)

designed, and tested intensively using focus groups and market research techniques. This particular ad was likely designed for a particular target market; the choice of Brad Pitt likely reflects assumptions about that target audience's age, race, class, and gender. The target audience (straight middle-class white women in their 30s and 40s, if I had to hazard a guess) presumably finds Pitt attractive and is therefore likely to be "persuaded" by the ad.

In practice, of course, the persuasiveness of the ad is much more complicated than that. Do you like scruffy Brad Pitt (pictured in the ad) or clean-shaven Brad Pitt better? Do you even find him attractive at all? Of course, one's interpretation of this ad after Pitt's divorce from Jolie-Pitt is different again: there can be no single, stable interpretation of this ad as persuasive or not. Are you team Brad or team Angelina? Do you still wish Pitt would just get back together with Jennifer Aniston? Do you believe the tabloid story that Pitt abused Maddox, the oldest child in the Jolie-Pitt brood? Does Pitt's intense gaze now connote something more sinister? Does the image evoke Pitt's adorkable jock character in *Burn After Reading,* or does the long hair make you think more of his more swoon-worthy character in *Legends of the Fall*?

Far from being stable, persuasive signifiers, then, celebrity advertisements are open to multiple responses that can only be considered stable and singular to the extent that a group of participants already selected for uniformity (in this case, age and gender) may *tend* to respond in certain ways. Further, ads of this type appear not out of thin air but in a context where users may expect to see this type of ad. Participants in this study would likely be familiar with this kind of advertisement from women's magazines, in which similar genres of ads appear. Most women's magazines feature similar kinds of ads that simply depict a product and a brand or logo: models wearing clothing from a high-end line, close-ups of a manicured hand wearing a particular color and brand of nail polish, and so on. The subgenre of women's magazine ad, then, tends to use less and less discursive rhetoric and more visual rhetoric, relying on viewers' familiarity with certain brands and their relative status. If you are Dolce & Gabbana or Chanel, you need do little more than evoke a particular image or mood with lighting, color, carefully selected clothing (what Project Runway judge and *Marie Claire* editor Nina Garcia would call "very editorial"), and carefully posed models. Sometimes, but not always, a few well-chosen words or lines of texts may make an explicit argument about the product.

The persuasiveness of such ads occurs not necessarily in a single moment of exposure but through repetition. One might see Brad Pitt and that bottle of Chanel No. 5 in several magazine ads, in a display at the makeup counter in Macy's (coupled with, perhaps, the smell of the perfume spritzed onto a card), on a bus, on a billboard, or in a television ad. Through the appeal of ubiquity, advertisers hope that people will build up a positive association for their product over time. Of course, the physical context through which people encounter this ad might differ. Consider, for instance, the different impact the ad might make as a display in a Bloomindale's store, amid bright lights, rows of makeup counters, and fashionable shoppers, for instance, as compared with the ad displayed on a bus stop on a crowded city street. In both cases, the ambience of the situation affects how the ad might be interpreted as well as its affect. Thus, the fMRI study by Klucharev, Smidts, and Fernández takes the idea of the celebrity advertisement out of context and seeks to distill it to a bare minimum (another example of neuroessentialism). In the process, the study ignores how people actually encounter such ads in real life.

HOW NEUROSCIENTISTS STUDY THE PERSUASIVENESS OF NEUROSCIENCE

Beginning in 2008, a conversation sparked within the field of neuroscience based on two key articles that took up the question of whether neuroscience— either texts referring to neuroscience or neuroscience images—had a persuasive effect. Researchers Deena Skolnick Weisberg and her colleagues published "The Seductive Allure of Neuroscience Explanations" in the *Journal of Cognitive Neuroscience* that year. In the study, Weisberg et al. provided participants with a series of sample texts similar to what you might read in a popular news article, some with neuroscience information added and some without. In their article, the authors report findings indicating that people rank scientific explanations as "more satisfying" when they include neuroscience information—even when the explanations themselves are "bad" or when the information in question is "logically irrelevant."[25] This claim has itself become a commonplace in discussions about neuroscience and its limits.

A mini-industry of critics took to blogs and Twitter accounts to identify instances of this "seductive allure" in popular and scientific articles alike. Books and articles aim to expose faulty scientific findings and lower our collective expectations about what neuroscience can tell us about our brains and ourselves. One such title clearly references the Weisberg study: Sally Satel and Scott O. Lilienfeld's *Brainwashed: The Seductive Appeal of Mindless Neuroscience.*[26] Often, these debunking efforts begin with the assumption that unsuspecting people are easily duped into believing junk neuroscience claims.

Yet, numerous articles have attempted to replicate the results that Weisberg et al. reported, with conflicting results. Two studies published in 2015 supported the original claims.[27] Other studies have yielded slightly different findings. For instance, Rebecca E. Rhodes, Fernando Rodriguez, and Priti Shah found that neuroscience information "increased ratings of scientist quality by 10% and improved mechanistic understanding by 26%" but did not affect the participants' ability to identify flaws in a study or to recall details of a study.[28] The authors conclude that neuroscience information "may be influencing participants' understanding about causation in the experiment" but that their study did not support the "seductive details" hypothesis.[29] Clearly, then, if neuroscience information does have an effect on readers, it is not yet clear exactly what that effect is.

Nonetheless, researchers have also been seduced by this "seductive allure" idea. At the time of this writing, the article has been cited 312 times, placing it among the top 1 percent of articles in the field of Neuroscience and Behavior according to the Science Citation Index. The phrase "seductive allure" has itself become a *topos* and article-naming convention in neuroscience discourse, both scientific and popular. Consider the evolution shown in these titles from scientific journals and popular science blogs:

25. Weisberg et al., "The Seductive Allure of Neuroscience Explanations," 472, 475.

26. Satel and Lilienfeld. See also Legrenzi and Umiltà, *Neuromania.*

27. Fernandez-Duque et al., "Superfluous Neuroscience Information Makes Explanations of Psychological Phenomena More Appealing"; Scurich and Shniderman, "The Selective Allure of Neuroscientific Explanations."

28. Rhodes, Rodriguez, and Shah, "Explaining the Alluring Influence of Neuroscience Information on Scientific Reasoning," 1437.

29. Rhodes, Rodriguez, and Shah, 1438.

- "The Selective Allure of Neuroscience Explanations"[30]
- "The Seductive Allure of Neuroskepticism"[31]
- "The Seductive Allure of 'Seductive Allure'"[32]

While the validity of the idea that neuroscience information is seductive is itself not established, *the idea itself can be said to be seductive.* That is, it seems somehow appropriate or fitting to believe that we are easily swayed by neuroscience information.

The "Seductive Allure" idea tends to be cited alongside another article that makes a similar, but not entirely parallel claim, and the two sometimes get conflated. Also in 2008, David P. McCabe and Alan D. Castel published "Seeing Is Believing: The Effect of Brain Images on Judgments of Scientific Reasoning." In this study, the authors asked participants to evaluate the "soundness of scientific reasoning" when viewing short articles that either did or did not include brain images;[33] they found that participants ranked the articles as more sound when they included images from fMRI scans than when they did not. Like the "Seductive Allure" article, "Seeing Is Believing" has been cited frequently (207 times at the time of this writing), which also ranks it in the top 1 percent of articles within its field. And like "Seductive Allure," it has been subject to replication attempts. In this case, however, researchers have mostly failed to reproduce the results. Researchers David Gruber and Jacob Dickerson have attempted the study with different kinds of images, including brain scans, stock photos, and graphs. This study failed to find a significant effect for any type of image. Similarly, Robert B. Michael et al. ran a series of ten replications of the original study, with over two thousand subjects, but found that brain images had little to no effect. The idea of the "influential brain image," the authors conclude, can be considered a "persistent meme" rather than a scientific reality. However, other researchers have found some effect for brain images that supports the original article, such as a 2011 article by Madeleine Keehner, Lisa Mayberry, and Martin H. Fischer that found readers to be more convinced when they included three-dimensional brain images as opposed to two-dimensional ones.[34] In this study, though, researchers did not include a condition with no image. In "Fooled by the Brain: Re-examining the Influence of Neuroimages," N. J. Schweitzer, D. A. Baker, and Evan F. Risko ran five different experiments, four that replicated McCabe and Castel's experiment. In those four, they found no evidence that neuroimages affected

---

30. Scurich and Shniderman.

31. Puglionesi, "The Seductive Allure of Neuroskepticism."

32. Farah and Hook, "The Seductive Allure of 'Seductive Allure.'"

33. McCabe and Castel, "Seeing Is Believing," 345.

34. Keehner, Mayberry, and Fischer, "Different Clues from Different Views."

readers. However, in a fifth experiment that compared neuroimages with other visual depictions of neuroscience data, they did find some effect. Thus, they conclude that neuroimages may be persuasive, but only relative to some situations.[35] In particular, "good explanations appear better after reading a bad explanation, explanations lacking a neuroscience basis seem less reasonable next to explanations containing such information, and realistic whole-brain images look more credible when compared to topographical maps of ambiguous brain activity."[36] Thus, neuroscientists have concluded that context matters—something that would not surprise a rhetorician.

Despite the lack of clear consensus among researchers, the idea that brain images are persuasive remains commonplace. In popular and scientific conversations about the persuasiveness of neuroscience, there exist two key ideas or topoi. One is the idea that (bad) neuroscience explanations are persuasive, and the other is that neuroscience images are persuasive. These ideas operate not as "strong arguments," then, but owe their persuasiveness to their fit with commonplace assumptions, beliefs, values, and emotions. For some scientists, the "persuasiveness of neuroscience" might seem compelling because it violates some of the core values of a scientific habitus: modesty, objectivity, correctness. If people can be persuaded by faulty neuroscience or by colorful brain images, then they are not being persuaded by objective, rational scientific evidence. This idea is threatening because it breaches scientific beliefs, and that threat might account for its visceral appeal.

For a rhetorician, it is difficult not to think back here to Gorgias's *Encomium of Helen*, which offers another take on the seductive power of language. Here, Gorgias offers four explanations for why Helen of Troy yielded to Paris. One of those explanations was that Helen may have "come under the influence of speech."[37] Notably, though, for Gorgias it is not only seductive nobleman wooing women who use persuasion; he also cites the examples of astronomers, "logically necessary debates," and "verbal disputes of philosophers" as instances of persuasion. Gorgias famously argues that the "effect of speech upon the condition of the soul is comparable to the power of drugs over the nature of bodies." For Gorgias, the distinction is just that, like drugs that produce different effects on the body, different types of language produce different effects on the soul: "some distress, others delight, some cause fear, other make the hearers bold, and some drug and bewitch the soul with a kind of evil persuasion."[38] For Gorgias, all language is seductive.

35. Schweitzer, Baker, and Risko, "Fooled by the Brain."
36. Baker et al., "Making Sense of Research on the Neuroimage Bias," 256.
37. Gorgias, *Encomium of Helen*, 52.
38. Gorgias, 53.

It may be, then, that what neuroscience researchers are worried about is not simply the "seductive allure" of neuroscience but, more specifically, the kind of "evil persuasion" that Gorgias refers to. After all, an article that might "seduce" readers into supporting neuroscience research or believing in the potential for new drug treatments would not be seen as nefarious. It is when readers are led to believe things that are not supported by principles of scientific reasoning or scientific consensus that the neuroscience researchers get concerned. As Weisberg et al. put it, "Logically irrelevant neuroscience information can be seductive—it can have much more of an impact on participants' judgments than it ought to."[39]

The concept of "seductive details" has proved useful for neuroscientists seeking to understand how neuroscience information can mislead readers. But in relying on that concept, they blur the boundaries between logic, comprehension, and persuasion. In fact, the "seductive details" concept developed out of an area of study concerned with "expository text processing," or comprehension of texts, not persuasion per se. In a 1989 article, Ruth Garner, Mark G. Gillingham, and C. Stephen White introduce the concept of "seductive details" in a study identifying how children and adults scored on tests of reading comprehension when extraneous details were added to sample texts.[40] They were not concerned with whether participants were *persuaded* by texts or not. Linking persuasion with the concept of "seductive details" thus implicitly assumes that persuasion mostly depends on logos. Yet, as we have seen, persuasion may also depend as much on affective responses, ethos, pathos, or emotion as it does with logic.

Indeed, a closer look at the studies in question reveals a confusion between persuasion, comprehension, and allure. Weisberg et al. asked their respondents to indicate how "satisfying" the explanations they provided were, which is a curious choice of term. One definition of "satisfy" is predominantly emotional, in the sense of fulfilling one's desires or appetites (as in a satisfying meal). Yet the term can also be understood in terms of logic; in mathematics, the term may be used to refer to fulfilling the requirements of a theorem. The term can also be used with reference to satisfying doubts or debts. It is unclear which of these understandings of satisfy participants were to use in the study.

In their interpretation of the results, however, the authors seem to assume the first definition, at least implicitly: "Explanations with neuroscience information . . . were rated as significantly more satisfying than explanations that did not include neuroscience information . . . Adding irrelevant neurosci-

39. Weisberg et al., 475.

40. Garner, Gillingham, and White, "Effects of 'Seductive Details' on Macroprocessing and Microprocessing in Adults and Children."

ence information thus somehow impairs people's baseline ability to make judgments about explanations."[41] Satisfaction with an explanation, here, is presumed to affect judgment by impairing reason and logic and somehow whetting readers' baser appetites. Here, emotion is considered an impediment to logic. A rhetorical view would suggest that satisfaction may have its basis in a range of other factors, including the intuitive appeal of brain-based explanations, the "fit" that such explanations seem to have with people's previous exposure to brain images and discourse (which are ubiquitous), or with the perceived *ethos* of neuroscientists as opposed to, say, psychologists.

In another test of the "seductive allure" hypothesis, Rebecca Rhodes, Fernando Rodriguez, and Priti Shah did not use the measure of "satisfaction" but instead used three measures: perceived article quality, perceived research quality, and convincingness. By separating out these qualities, they found that they were not able to support the "seductive details" hypothesis—that is, the neuroscience information increased ratings of both article quality and research quality, but not of convincingness.[42] Thus, it seems that adding neuroscience details boosted the ethos or credibility of the articles without making them more persuasive. Here, the neuroscience details did not affect participants' ability to identify flaws in the studies described or their ability to recall details of the study (which is a truer measure of the "seductive details" theory as originally described by Garner, Gillingham, and White). Comprehension or recall of information, in short, is different from persuasion. Seductive details may lead one astray—they may confuse the issue or prevent readers from grasping content—but they do not necessarily make a popular science article more convincing.

A NEURORHETORICAL APPROACH TO PERSUASION

Across all these studies, we see the following assumptions being made about persuasion:

- Persuasion can be quantified / measured on a scale
- Persuasion happens at a specific moment
- Persuasion can be broken down into component parts or functions
- Component parts of persuasion can then be mapped onto the brain
- Persuasion can be studied in a laboratory or relative vacuum; context does not matter

---

41. Garner, Gillingham, and White, 472.
42. Rhodes, Rodriguez, and Shah.

- Persuasion can be studied in a small homogeneous group of people (n = 20 or so) but the results can be applied to humans at large
- The specific content of actual arguments or messages does not matter (or matters little), as long as they can be similarly rated or grouped (i.e., as highly persuasive or highly unpersuasive) or as positive/negative, etc.

The limitations of this approach become apparent when neuroscientists turn their attention to their own discourse. Indeed, neuroscientists worry about the persuasiveness of neuroscience itself—including images of brains that circulate in popular accounts and popular news articles reporting the latest findings. Yet their attempts to study this phenomenon are limited by their assumptions about what persuasion is and how best to study it.

To overcome these limitations, I will show how the method I developed in the previous two chapters can be integrated as a heuristic to analyze existing approaches to the study of persuasion and to identify alternatives. The heuristic shown in Table 1 identifies three different tasks: raveling back, raveling out, and raveling together. Raveling back includes two actions: identifying the metaphors that inform an area of study and then questioning definitions (which can help to identify the limitations of neurorealism). Raveling out involves puzzling through questions related to the study team and study design (which addresses challenges of neuroessentialism). And raveling together involves bringing together multiple disciplinary perspectives to generate alternative interpretations of a study that can help to redress some of the limitations of neurorealism and neuroessentialism.

TABLE 1. A Heuristic for Neurorhetorical Analysis

TASK	ACTIONS	QUESTIONS
Ravel Back	Identify Guiding Metaphors	What guiding metaphors are currently at play? For what do they *select* and from what do they *deflect* our attention? What other metaphors are possible? How might different metaphors open up different lines of inquiry?
	Tease Out Definitions	How are we defining the concepts in question? What other definitions are possible?
Ravel Out	Consider Study Team	Who else might know about this concept? What other literature might we consult?
	Consider Study Design	What other hypotheses might we generate? How else could we perform this study? What other variables may be relevant?
Ravel Together	Resituate the Study	What other literature or theory can we cite in our introduction or discussion? How can this literature add texture to our study findings? What limitations does this literature point to?

A HEURISTIC FOR TRANSDISCIPLINARY RESEARCH IN SOCIAL AND COGNITIVE NEUROSCIENCE

Raveling Back

In the studies of persuasion examined here, we see three different metaphors guiding the discussion: that of the "strong" argument, the "seductive" argument, and the "satisfying" argument. Each metaphor entails different assumptions about how we are persuaded. In the first case, calling an argument "strong" implicitly imagines arguments as bodies or as materials (like, say, lumber or steel) that can be categorized according to their strength. Strength is usually defined in terms of how well the argument draws on facts, logic, and, to some extent, emotional appeals. A sufficiently strong argument should be persuasive to most people. The assumption is that strong arguments are good arguments. The metaphor of seduction, in contrast, defines persuasion in a more clearly affective manner, suggesting that people can be misled predominantly by emotional appeals in the same way that Helen was seduced by Paris. That is, people might ignore a logical argument in favor of one that is emotionally compelling. And the metaphor of "satisfaction" entails a slightly different perspective—one might call it an appetitive metaphor—in which people are sated, or satisfied, by arguments in the same way that a hungry person can be sated by a candy bar. Here again, the idea is that the argument appeals to certain kinds of predilections, but again in ways that are potentially misleading. The "strong" argument metaphor holds out some suggestion that persuasion can be a good thing, but the latter two suggest a suspicion of persuasion that can of course be traced back to ancient times (for instance, to Plato's famous alignment between rhetoric and the arts of flattery in the *Gorgias*).

Within scientific disciplines, we might locate this suspicion more specifically in the history of scientific discourse itself. In 1667 Thomas Sprat published a *History of the Royal Society of London*. In a section of that text that has been often anthologized, Sprat warns his fellow Society members against the powers of rhetoric: "eloquence ought to be banished out of all civil societies as a thing fatal to peace and good manners."[43] Rhetorical figures, in particular, "give the mind a motion too changeable and bewitching to consist with right practice"; they represent a "beautiful deceit" that is dangerous for those focused on the pursuit of knowledge about the natural world.[44] Later, Sprat refers to rhetoric as an "infection."[45] We see echoes of Sprat's warning

---

43. Sprat, *The History of the Royal Society of London*, 2.
44. Sprat, 2.
45. Sprat, 4.

in the phrase "seductive allure" used in neuroscience studies. Like Sprat, the authors of these studies seem to believe that discourse (images or texts) about the brain can be "seductive," can lead people astray (to take the etymological definition of the term "seduce"). Brain images, like rhetorical figures, may be beautiful, but they can also dazzle readers and lead them away from proper thinking. These assumptions are embedded in study designs that seek to know not whether neuroscience explanations or images are persuasive, per se, but whether they can trick people into believing things that are not true.

Next, we might compare these metaphors and definitions with those used in other fields, namely rhetoric. However, no single conception of persuasion would be universally accepted within the field of rhetoric. Instead, metaphors or definitions of rhetoric might lead us to identify specific elements of persuasion that one might wish to study, such as identification (in a Burkean model), the effect of visual images, the effect of repeated exposure to arguments over time, the effect of specific contexts for viewing an argument, and so on. As I have suggested above, these models of persuasion would also point to the circulation and accretion of argument over time, or to contextual factors, as I will explain in the next section.

Raveling Out

Given the above, researchers studying persuasion in a neuroscientific setting might, quite obviously, be interested in including a rhetorician on their team. Then, they might reconsider how the study design might better reflect a specific aspect of persuasion. For instance, they might design a study that includes contextual factors, longitudinal studies of exposure to a set of arguments over time, or supplementary methods such as interviews about pervious experiences.

For instance, it is worth nothing that studies of persuasion in neuroscience based on experimental models typically focus on just one genre, the popular article reporting one finding. While that is indeed one way that people are exposed to neuroscience, it is certainly not the only one—and may not be the most common one. One remarkable aspect of neuroscience discourse is that it has permeated a range of genres, not just those dedicated to science news. It is not uncommon to find neuroscience information creep into parenting magazines, news articles on educational policies, television shows, or even the aisles of your grocery store (see, for instance, the line of drinks marketed under the name "Neuro"). It is this permeation of neuroscience in public life that Thornton and Rose refer to in their work as a more general cultural discourse. These instances of neuroscience discourse assert themselves not necessarily as single instances of persuasion but as part of a "body of identifications." Thus, focusing on only one genre within a laboratory study (typically news articles) limits a deeper understanding of how these claims work within different genres and collectively, across genres. One might imagine a targeted study that seeks to identify whether and how different genres matter in people's responses to neuroscience arguments.

Alternatively, one might imagine studies in which neuroscience were introduced in different contexts (comparing their use in, say, parenting magazines versus television shows), or in repeated exposures over time (rather than a single exposure as in most fMRI studies). Do neuroscience explanations stick with people a week or a month later? If allowed to explain their reasoning for why a particular explanation is compelling (or not), what more can we learn about the persuasiveness of neuroscience arguments?

Raveling Together

Finally, given the rich set of sources about persuasion that rhetoricians carry with them, researchers might work together to integrate knowledge from each contributing discipline to the literature review, analysis, and discussion sections of any resulting publications. While neuroscience studies have not found much evidence to support the idea that neuroscience is persuasive, these studies have been limited by the same methodological assumptions I outlined above, which assume that persuasion occurs in a single moment that can be artificially induced. It is harder to interrogate whether and how neuroscience details in a text (or images) function in the broader sense of *persuasions* that accrete over time. Neuroscience information may be doing something other than rationally persuading us. To return to the Brad Pitt example above—or to the more general topic of celebrity endorsement advertisements—it may not necessarily be a single exposure to a perfume ad in a magazine that persuades someone. What renders the ad persuasive is the circulation not just of that ad but of the celebrity and his or her image in our culture—the "dull daily reinforcement" of his or her celebrity.[46] Similarly, the persuasiveness of messages about neuroscience itself depends on the repetition and circulation of those messages in public service announcements, magazine advertisements, and so on, their connections to other arguments and even material items, such as energy drinks, brain apps and games, books, and toys that are marketed using the language of neuroscience.

---

46. Burke, *Rhetoric*, 26.

In the scientific model used to study persuasion in neuroscience, we have a much simpler model, one that more closely tries to resemble a sender-message-receiver model of communication that rhetoricians recognize as woefully inadequate. As Jenny Edbauer Rice argues, "rhetorical processes operate within a viral economy. The intensity, force, and circulatory range of a rhetoric are always expanding through the mutations and new exposures attached to that given rhetoric, much like a virus."[47] Rice emphasizes that these viral exposures are *affective,* not simply occurring on the level of conscious, rational thought.

It is one thing to inquire into the potential for neuroscience information to mislead audiences, which is the perspective taken in most of the studies undertaken by neuroscientists in the "seductive allure" paradigm. I would argue, though, following Nikolas Rose, Davi Johnson Thornton, and others, that neuroscience discourse (images and texts) circulates in ways that do more than persuade us of the truth or falsity of a particular finding or claim. When considering neuroscience discourse more broadly, as a series of "viral exposures," we can understand it as conditioning readers to brain-based ways of thinking about ourselves and our behaviors—in fact, these discourses constitute readers as such.

Thornton argues that the brain represents a "fixation that extends far beyond the spheres of science and medicine and infiltrates virtually every corner of daily existence."[48] Yet, she posits, the meaning of these images is not fixed but depends on how they are "framed, contextualized, and used in different settings."[49] Images, in particular, assert "the ethos of scientific authority" and carry a sort of truth value, even if that truth is not fixed or set. For instance, the same brain image might be used in one popular article to argue for the essential differences between men's and women's brains and in another article to argue against that essential difference. In the aggregate, though, we might say that the circulation of neuroscience discourse and images does tend to produce certain effects. For instance, Thornton continues, these images and texts "purport to show the biological basis of all aspects of experience," and they present an argument for the importance of science (and of neuroscience in particular) in shaping how we understand our lives.[50] They may also tend to constitute what counts as a normal or healthy brain and, in the process, may reproduce understandings of different brains as abnormal, ill, or diseased. Yet the "normal" or "healthy" brain is also presented as something that can always be optimized through brain training programs.[51] In studies that simply present

47. Edbauer, "Unframing," 17.
48. Thornton, *Brain Culture,* 2.
49. Thornton, 4.
50. Thornton, 12.
51. Thornton, 22.

participants with a statement and ask them to indicate whether they were convinced or not, it is difficult to factor in their previous "viral exposures" or to separate them from the exposure in the experiment itself. Researchers might at least acknowledge this limitation when describing research that assumes a "single shot" exposure to a neuroscience argument.

Thus far, then, I have brought the tools of rhetorical theory to bear on social neuroscience, suggesting a heuristic method that can help to inform what I am calling "neurorhetorical" studies. In the next three chapters, I will apply this framework to three case studies that are of interest to rhetoric scholars and neuroscientists alike: sex/gender, politics, and affect. In particular, I will highlight research that is already moving toward a recognition of contingency, context, and the interplay of culture, language, and the brain. In these three areas of research, we find some potential for a type of study that would ravel together neuroscience and humanistic study. In what follows, I identify areas where elements of the heuristic shown in Table 1 are already in play, and I identify further areas for collaboration across disciplinary perspectives.

CHAPTER 4

Neurosex

ACCORDING TO a *Guardian* article published in 2018, "provocative clothes" have an effect on the brain. And while that effect is "no excuse for assault," it is still one that has "many cognitive components" worth considering:

> We observe something, our prefrontal cortex—via links to the more funda-mental emotional and reward systems—analyses it and determines if it's sex-ual in nature, and if so, if it is "sufficiently" sexual (eg we find some people sexy, but not others). If it is, our attention is directed towards it, and emo-tional and motivation processes are activated via our amygdala and anterior cingulate cortex respectively. It's incredibly complex in detail, but the neuro-logical systems that regulate arousal and desire do indeed have many potent effects via important regions throughout our brains.

The "we," of course, refers predominantly to men, who, as the article notes, are "far more visual in nature" than women, especially when it comes to sexual arousal. Not to worry, though, the article goes on to say, because the brain may be implicated in arousal but also has handy parts that can regulate or sup-press sexual behavior, such as the prefrontal cortex.[1]

---

1. Burnett, "How 'Provocative Clothes' Affect the Brain."

The content of this article is, clearly, bunk. That it was published in 2018 is, sadly, not all that surprising. And the rhetorical workings of this type of claptrap have been widely exposed, as I will briefly describe below. In this chapter, though, I demonstrate how at least some neuroscience researchers are moving toward a kind of approach more amenable to engagement with humanities researchers. By questioning gender stereotypes and how they emerge in brain research, these scientists open up possibilities for "raveling together" humanistic and scientific insights. To identify how alternative kinds of research can be possible, we must turn our attention to the experiment as the unit of analysis. Only by understanding how scientific knowledge is produced through specific experimental entanglements can we begin to identify new possibilities that better capture the interaction of the sociocultural and neurobiological networks through which brain sex/gender differences may (or may not) emerge.

In this chapter, I attend to the metaphors and definitions guiding sex, gender, and sexuality research in social and cognitive neuroscience that uses fMRI, consider study designs that enact those definitions, and identify possibilities for a "neurorhetorical" approach. I begin by briefly outlining how sex and gender are typically operationalized in social and cognitive neuroscience research. Second, I develop a framework for examining the entanglement of sex/gender within neuroscience research by closely analyzing scientific studies that suggest how research methods may either reify sex/gender differences as biological or open up inquiry into the rhetorical and social production of those differences, considering how those differences are generated in society at large and within the context of the experiment itself. I show how this framework can be applied to neuroscientific imaging studies of sex/gender difference, showing in particular how some recent neuroscience research actually supports the viewpoints of feminist researchers who understand the sociocultural and neurobiological to be intertwined. Third, I consider how this framework can be applied to research that focuses more specifically on sexuality. This research tends to disavow sociocultural and rhetorical considerations, and in the fourth section I show how paying attention to those considerations could enrich our understanding of human sexuality and how the neurological elements of sexuality are entwined with sociorhetorical factors. Rather than arguing for a single way of understanding sex/gender, such an approach might take the position that gender and sex are "multiple ontologies"—that is, what sex or gender *is* depends on how it is enacted in a particular material time and location, in this case, that of the experiment. The question, from this perspective, becomes not what the true essence of sex/gender is, but what kind of research is needed, and for what purposes.

HOW NEUROSCIENTISTS STUDY SEX AND GENDER

Researchers interested in sex/gender differences in neuroscience often begin from an antithetical commonplace about men and women (such as the idea that men are more interested in sports than women are) and then seek to explain it by blending psychological and neuroscientific methods. The dominant rhetorical figure is not a metaphor, per se, but instead synecdoche and antithesis. Such studies often begin by assuming that male and female brains are separate and distinct types epitomized as the "female brain" or the "male brain." Here, the brain is imagined not in terms of something that it lacks (as in a metaphor) but by the substitution of a part for the whole: the (male or female) brain for the whole person.

These gendered brains are then defined in terms of specific capacities presumed to be distinct in men and women. For instance, researchers might measure spatial ability in male and female participants using an object rotation task while they are in an fMRI machine; they would then determine, statistically, how well men and women did on the task and identify which brain regions men and women used while performing it. The starting assumption they made—that men and women have different spatial reasoning abilities—shapes the materials of the experiment itself, including the questionnaire used and the brain regions examined in the fMRI machine. Of course, by raveling back, we could connect these assumptions to long-standing beliefs about women. Historically, for instance, it was assumed that women's smaller brain sizes and weights indicated that they were unsuited for advanced academic study or for playing key roles in public life. In 1869, for instance, J. McGrigor Allen wrote that women's brains, given "the small and elongated cerebel, or organ of will," were more oriented toward feeling and emotion than to "exertion, mental or bodily" (where men, of course, excelled).[2] Today, fMRI-based research often maintains this fundamental rhetorical figure of antithesis, albeit focusing less on sheer size or shape of brains and more on the relative effects of BOLD activations in particular regions.

Typically, these articles begin with generalizations and then seek to study them experimentally:

- "Behavioral studies suggest that females often perform better in emotional tasks than males."[3]

2. Allan, "On the Real Differences in the Minds of Men and Women," cciii–cciv. See also Vidal, "Brain, Sex, and Ideology" for a review of this type of discourse.

3. Schulte-Rüther et al., "Gender Differences in Brain Networks Supporting Empathy," 393.

- "In the past few decades, scientists and the public alike have debated the existence of gender differences in mathematical skills."[4]
- "Men and women . . . differ markedly in aspects of sexual behavior, such as the reportedly greater male interest in and response to sexually arousing visual stimuli."[5]
- "Men outperform women on several spatial ability measures."[6]
- Disorders such as depression and posttraumatic stress disorder "have a substantially greater prevalence (1.5–3 times) in females than males."[7]

These sentences tend to belong to the category that Bruno Latour refers to as positive modalities, or sentences that "lead a statement away from its conditions of production, making it solid enough to render some other consequences necessary."[8] These sentences effectively "black box" the statement, making it seem unassailable, "an indisputable assertion" that is typically backed by at least one reference to previous research that is being marshaled to support the black-boxing of the statement.[9] In short, each of these sentences offers a fact-based argument, the first stasis point. We might summarize the arguments in these first lines as follows:

- Women are more empathetic than men.
- Men are better at math than women.
- Men like porn.
- Women can't read maps.
- Women are more emotional than men.

While some of these studies couch those stereotypes in results from behavioral or psychometric research, they nonetheless take a stereotype or a generalization at face value and then seek to identify its neural correlates. These studies tend to rely on neurorealism and neuroessentialism, assuming both that sex/gender distinctions are real and that men's and women's abilities and interests can be essentialized.

---

4. Keller and Menon, "Gender Differences in the Functional and Structural Neuroanatomy of Mathematical Cognition," 342.

5. Hamann et al., "Men and Women Differ in Amygdala Response to Visual Sexual Stimuli," 411.

6. Seurinck et al., "Does Egocentric Mental Rotation Elicit Sex Differences?" 1440.

7. Williams et al., "Distinct Amygdala–Autonomic Arousal Profiles in Response to Fear Signals in Healthy Males and Females," 618.

8. Latour, *Science in Action,* 23.

9. Latour, 23.

Scholars have identified how these assumptions play out in brain research. In a now-canonical article, "How Bad Science Stays That Way," Celeste Condit (1996) argues that brain sex research relies upon "commonsense understandings of the nature of the dispute about male and female sex and gender," leading to "bad science"—science that produces accounts that are "insufficiently rich to account for the material phenomenon under investigation."[10] Condit points to neuroscience research that leads toward neurorealism and neuroessentialism—research that either assumes that sex/gender brain differences are real and concrete, or research that tends to essentialize what counts as a "male" or "female" characteristic or quality.

Since then, the growth of neuroscience has led to more and more accounts that reproduce the tendencies Condit noted, relying on commonplaces (or *topoi*) about men's superior map-reading skills or women's interest in shopping. Other researchers in rhetoric have also pointed out these tendencies in brain research. Jeanne Fahnestock shows how neuroimaging research produces antithetical visual and verbal arguments by exaggerating differences between men and women, using the rhetorical figure of antithesis.[11] In Anne Fausto-Sterling's terms, this type of research mistakes sex/gender differences for sex *dimorphism*, rhetorically pushing men and women apart by exaggerating what are, in fact, small *average* differences between men and women.[12] More recently, Christa Teston has called attention to how the material practices of neuroscience research, especially null-hypothesis significance testing (NHST), can lead researchers to overemphasize sex/gender differences that have statistical significance but that are relatively small in effect size. Here, a minor difference between male and female participants on a given measure is translated as a major sex difference, especially in popular news accounts.[13]

With relation to neuroscience research, in particular, a slate of recent books has thoroughly dismantled results that purport to identify measurable anatomical or connective differences in male and female brains. Cordelia Fine's *Delusions of Gender* and Rebecca Jordan-Young's *Brain Storm* are both excellent resources for those seeking to understand that research. To date, then, the predominant position among feminist scholars has been to critique brain research, to point out its flawed reasoning and to lambaste its tendency to uphold antiquated gender stereotypes. This fallback position, though, poses problems because it begins from a standpoint of critique rather than engagement. Recently, feminist science studies scholars have called for the latter

---

10. Condit, "How Bad Science Stays That Way," 88, 87.
11. Fahnestock, *Rhetorical Figures in Science*.
12. Fausto-Sterling, *Sexing*.
13. Teston, "Enthymematic Elasticity in the Biomedical Backstage."

position. Karen Barad, for instance, argues that feminist researchers should avoid the traditional position in science studies, which has been to "position oneself at some remove, to reflect on the nature of scientific practice as a spectator," and usually, to launch a critique of those practices.[14] However, researchers who make a different rhetorical decision might design a study differently, and in this section I "ravel out" by examining studies that take an alternative approaches to the study of sex differences in the brain are possible.

Next, I will ravel out (or puzzle out) questions of study design not by critiquing what Condit would call "bad science," but by examining studies that attempt to study sex and gender differently. These studies will show how starting from a different assumption leads researchers to pay attention to different aspects of the experiment to materialize gender differently. For one, rather than beginning with an assumption of difference between men and women, researchers can investigate how gendered stereotypes work. One study, "Neural Correlates of Stereotype Application," by Jason P. Mitchell et al., begins by outlining the broad phenomenon by which people mentalize about others by activating stereotypes. The researchers explain that stereotypes guide peoples' judgments about others' psychological characteristics and behaviors.[15] For example, the researchers showed participants a picture of a man or woman and then asked them to judge whether a statement would apply to that person or not, such as "likes scented candles" or "enjoys watching football."[16] In rhetorical terms, the researchers prompted people with commonplaces about what men and women are supposed to like—with essentializing statements.

The authors suggest that we may draw upon the right prefrontal cortex, an area of the brain associated with categorization, when making stereotyped judgments. Those participants who scored higher on measures of gendered attitudes, in fact, showed greater activations in that region of the brain. The authors conclude that when applying gendered stereotypes to individuals, we draw on a more fundamental cognitive process of categorization. In Burkean terms, we might term this a sense of "what properly goes with what," or a piety that develops over time through repeated exposure to stereotypes.[17] This research, then, points to a potential example of what Fausto-Sterling and Barad term "dynamic systems" or "entanglements," respectively, or the mutual imbrication of neurological functions with social and rhetorical constructs of sex/gender.[18] Of course, we need to take this finding with the same level of

---

14. Barad, *Meeting the Universe Halfway*, 247.
15. Mitchell et al., "Neural Correlates of Stereotype Application," 594.
16. Mitchell et al., 595.
17. Burke, *Permanence*, 74.
18. See Barad, Fausto-Sterling, Coll, and Lamarre, "Sexing the Baby."

skepticism that we have employed throughout; this study relies on the same techniques for producing difference, including NHST and the assumption that blood oxygenation levels are a proxy for brain activity, that have been critiqued elsewhere.[19] It does suggest, though, that there are different ways of working within the dominant paradigm in fMRI social and cognitive neuroscience research to ask different kinds of questions.

We see similar a similar approach in a study by Susanne Quadflieg et al., "Exploring the Neural Correlates of Social Stereotyping." In this study, participants were asked to indicate whether certain kinds of activities, such as mowing the lawn, watching talk shows, or taking photographs, corresponded more with men or women (or neither). While they were doing this task, researchers monitored their blood oxygenation-level-dependent activation in an fMRI. The researchers correlated those activation responses with participants' results on two questionnaires, the Implicit Association Test and the Attitudes Toward Women Scale. They found patterns of activation in regions of the brain associated with activity knowledge and with evaluative or emotional processing (the amygdala). An interesting implication from this study is that gender stereotyping may be fundamentally evaluative, but that it may also depend on learned associations.[20] In fact, those who scored highest on the scales used to test gendered beliefs had the greatest activation in the right amygdala, a region of the brain associated with "stimuli that have acquired emotional significance through learning rather than based on innate propensity." In other words, the authors propose that while stereotyping is a nearly universal human capacity, "activity in the neural circuitry supporting these responses is sensitive to the strength with which gender-based beliefs are endorsed in everyday life."[21] This is a provocative suggestion. It suggests that, rather than viewing sex/gender differences as hardwired or innate (as many seem to do), scientists might now consider those differences to be based on a fundamental cognitive capacity to make judgments, coupled with a learned capacity to make those judgments based on previous exposure to stereotypes. Here, the neurobiological and sociocultural are potentially entangled, with learned gender stereotypes linked to activations in the right amygdala.

Here again, we must be mindful of how this study still participates in the traditions of neurorealism and neuroessentialism that I outlined in the first part of this book. Does this study simply move "stereotyping" into the place of sex/gender difference, making stereotyping "uncritically real" and essentializing how stereotyping functions by operationalizing it in this manner? Possibly.

---

19. See Teston.

20. Quadflieg et al., "Exploring the Neural Correlates of Social Stereotyping," 1567.

21. Quadflieg et al., 1567.

However, my argument here is that there is, currently, no way for neuroscientists to produce acceptable knowledge using fMRI without drawing on this rhetorical infrastructure in some way. Thus, I am suggesting that the Mitchell et al. and Quadflieg et al. studies are simply working differently within that framework, starting from a different assumption to ask different questions. Their results, further, should be taken not as "proof" that stereotyping works in this manner but as suggesting different possibilities for thinking through sex/gender as a rhetorical system that enacts material differences.

Two other studies suggest additional ways that researchers may work differently within the rhetorical infrastructure of neuroscience by taking the very act of ascribing sex/gender to individuals as the object of study. In the first example, "Unification of Speaker and Meaning in Language Comprehension: An FMRI Study," Cathelijne Tesink et al. investigated how individuals make judgments about an unknown or invisible speaker's age, sex, and social background by interpreting their voice. To examine this phenomenon, the researchers recorded sentences pronounced by male or female speakers, children or adults, and, in the final condition, speakers with an accent associated with either an upper-class or lower-class accent (in Dutch). Some of these sentences conveyed information that would be expected of the speaker, and others did not. For instance, the incongruent utterances included a male speaker saying "My favorite colors are pink and lime green" or a child saying "Every evening I drink a glass of wine before going to bed."[22] These sentences violated common social stereotypes.

While the authors do not speculate about what their findings mean with relation to gender stereotyping (or other forms of stereotyping) in particular, they did find activations in the participants' brains when participants heard incongruent sentences that suggested "increased and more prolonged efforts to search and retrieve semantic knowledge about the likelihood of events occurring in the real world."[23] These activations occurred in the left inferior frontal cortex, an area of the brain associated with language comprehension. The authors further suggest that the process of "unification" also entails other kinds of information—not just words, but gestures and other extralinguistic vectors of information. Thus, this unifying process involves a variety of inputs, linguistic and otherwise. The study indicates, then, that gender (and other) stereotypes are partially enacted through discursive constructs, and that we store semantic knowledge in a manner akin to what Burke calls piety. By taking sex/gender stereotyping not as a given but as itself the object of investi-

22. Tesink et al., "Unification of Speaker and Meaning in Language Comprehension," 2086.
23. Tesink et al., 2097.

gation, researchers complicate the simple binary that shapes so much of the "bad science" that Condit decried in her article. Instead, we again see how the entanglement of sociocultural and neurobiological elements can itself be studied scientifically.

A different approach involves questioning the process of gender identification, but this time in relation to how we ascribe gender to human motion—specifically gait. The study asked how it is that we ascribe gender to an individual's walk. To answer this question, the researchers attached lights to the major joints of an actor's body. The actor then portrayed male, female, or gender-ambiguous walks and asked participants viewing only those lights to identify the gender of the walker. Importantly, the researchers found that this discrimination of gender is contextual. That is, after viewing a "female" walk, viewers tended to identify the next stimulus as a "male" walk, and vice versa. The authors conclude that "gender identification of human walkers is rapidly malleable and subject to adaptation."[24] In other words, while gender identification may be a neurological function, these researchers suggest that it is by no means stable but is intertwined with the rhetorical context. Here again, it is important to acknowledge the limitations of the study as well. How, exactly, did the actor portray gendered walking behavior? Clearly, to do so, he or she must have drawn on available gender stereotypes to stand in for a male or female walk, and those stereotypes may have been what was being observed in the study, not a "female" or "male" walk per se. Nonetheless, this study offers an intriguing heuristic insofar as it opens up possibilities for seeing gender identification of others as malleable.

Together, this set of studies suggests that researchers who do not take male/female differences as a given can enact sex/gender differently, challenging the antithetical reasoning foundational to so many other studies. They also refuse the polarization between sociocultural and neurobiological elements, showing how they are also entwined. While by no means the most common approach among fMRI studies of sex/gender, these examples indicate that brain research can interrogate links between the rhetorical, cultural, and biological rather than taking sex/gender binaries as natural and obvious.

More recently, Daphna Joel and her colleagues have published a series of intriguing articles that offer a new metaphor for understanding the relationships between female and male brains. Rather than assuming a dichotomy or antithesis, Joel and her colleagues argue that "most brains are comprised of unique 'mosaics' of features, some more common in females compared

---

24. Jordan, Fallah, and Stoner, "Adaptation of Gender Derived from Biological Motion," 739.

with males, some more common in males compared with females, and some common in both females and males."[25] In a meta-analysis of brain research, Joel et al. found that the vast majority of brains demonstrate "internal inconsistency"—that is, very few brains (0.1 percent to 1.8 percent) possess only characteristics associated with the biological sex of their body. Based on this analysis, Joel et al. call for a "paradigm shift" in sex difference research, an approach that would allow for heterogeneity and differences in the "human brain mosaic."[26]

Clearly, Joel's call for an alternative approach poses challenges for researchers who are used to relying on the male/female antithesis in their research designs. I would argue that the elements I outlined above go part of the way to addressing this challenge. First, researchers might focus less on male/female differences—especially if, as Joel puts it, *there are no 'true' 'male' and 'female' brains out there to discover.*[27] Instead, they might focus on how individuals ascribe sex/gender differences to others and how exposure to stereotypes shapes gendered expectations. They might also continue to examine the entanglement of the social and rhetorical with the biological. How is it that individual brain "mosaics" come to be? How do individuals, as Janet T. Spence argues, "use their gender-congruent characteristics to verify and maintain their gender identity,"[28] even as they embrace other characteristics that are not gender-congruent—and what neurological traces do we find for those choices? Such an approach would offer an opening for scholars trained in the humanities—especially in feminist, gender, sexuality, and women's studies—to suggest innovative research designs.

However, the research examined here still largely draws on dated, binary models of sex and gender. While these studies investigate the extent to which gender differences are produced, few studies take up a perspective of gender as polymorphic or diverse. Most studies assume participants are either male or female, masculine or feminine. The Tesink et al. study included an "ambiguous" case, with the assumption being that the walker had a true gender but that it was not identifiable from the person's walk. The study designs do not allow for a more complex model that would include gender diversity—that is, a wider range of sexed and gendered identities, including transgender, genderqueer, or agender. Even Joel's theory ignores individuals who identify in ways other than male/female or masculine/feminine. Including a wider range of individuals in sex/gender research, but without pathologizing those differ-

25. Joel et al., "Sex beyond the Genitalia: The Human Brain Mosaic," 15468.
26. Joel et al., 15472.
27. Joel, "Male or Female? Brains Are Intersex," 4; emphasis in original.
28. Spence, "Gender-Related Traits and Gender Ideology," 633.

ences, remains a significant challenge for researchers in the neurosciences. I will "ravel together" humanistic theories of sex/gender with neuroscience ones in the conclusion of this chapter.

HOW NEUROSCIENTISTS STUDY SEXUALITY

So far I have considered research that relates primarily to sex/gender and the brain, but similar challenges face research into human sexuality and the brain. We are now schooled to understand the brain as the most important "sex organ," a view that both popular news outlets and scientific ones espouse. Popular writer Daniel G. Amen asserts that "even though it feels genital, the vast majority of love and sex occurs in the brain," which is "the largest sex organ in the body."[29] Recognizing the significance of the brain, according to Amen, can enhance your sex life. Feminist writer Naomi Wolf posits a direct brain–vagina connection in *Vagina: A Cultural History*, while psychologist William M. Struthers warns that pornography can "hijack a man's brain, hypnotizing him and rendering him incapable of making good decisions."[30] The metaphor of the brain as a "sex organ" seems to have emerged in popular and scientific discourse in the 1980s and 1990s,[31] but it now shapes how researchers study sexuality, positioning the brain as the location of sexual response and therefore as the proper place in which to investigate sexuality.

Isabelle Dussauge has expertly raveled out the methodological assumptions shaping such research in her article, "The Experimental Neuro-Framing of Sexuality." Based on her review of studies of sexuality, sexual behavior, and desire, she finds the following tendencies. First, the studies begin by choosing "ideal subjects," typically highly responsive, young (between 18 and 40), and most often heterosexual males.[32] Second, Dussauge explains, a heterosexual orientation is taken as the norm in these studies, and participants are most often selected based on their response to a Kinsey scale. Those who report same-sex desire are excluded in most studies. Despite this exclusion, though, the results of the study are either taken to represent sexuality writ large (not sexuality in that particular group) or sexuality of group A (say, men) in com-

29. Amen, *The Brain in Love*, 1.

30. Wolf, *Vagina*, 3; Struthers, *Wired for Intimacy*, 11.

31. The earliest instance of this metaphor I could find is from 1981 in Silber, *The Male: From Infancy to Old Age*, 32. Further instances appeared in the early 1980s in books providing advice to women and men regarding their sex lives, especially in discussions surrounding impotence. However, the term also informed how sexual identity was understood. For an analysis of how this trope was deployed in the famous John/Joan case study, see Sloop, *Disciplining Gender*, 43.

32. Dussauge, "The Experimental Neuro-Framing of Sexuality," 126.

parison to group B (say, women). Any other study group that may be chosen by researchers (homosexual men, lesbians, people with disabilities, etc.) is marked by their deviance from the default, and the findings do not tell us about sexuality, in general, but about sexuality in that particular group.

Dussauge also explains how these neuroscience studies define and operationalize sexual desire in ways that are meant to evoke a predictable response at a predictable level of intensity. For example, subjects might view pornographic images preselected for their appeal to a test group, with the goal of ensuring that participants in the study will respond similarly to the stimulus. In this way, Dussauge argues, these studies assume "that there is a universal desire and pleasure which, once triggered, is the same for everyone" and that leads to a predictable sequence of events (desire, arousal, orgasm).[33]

Often what counts as sexual response is defined in part by a set of control stimuli (often images of bodies engaged in sports) that presumably do not evoke desire, despite the fact that athletes are routinely portrayed for us in ways that evoke, at least, admiration of their bodies, if not sexual desire. (ESPN, for instance, regularly produces a magazine issue that features athletes naked. Female athletes are, of course, especially likely to depicted in nude or suggestive poses.) One might also point out that in these studies, strong sexual response is taken as the norm and ideal. Individuals who identify as asexual, or with one of the many asexual subidentities (such as demisexual, semisexual, or gray asexual), are simply not included in these studies. In this way, the lack of a sexual response is by default portrayed as abnormal.

Finally, Dussauge describes how neuroscience studies eliminate the body, locating an idealized desire in the brain: a "pure (ageless), perfectly oriented along the homo/hetero-axis, bodiless, distillated to an essence independent of its objects and feelers."[34] In other words, despite the decidedly distributed, embodied response that sexual desire and pleasure evoke, that wider range of response is stripped out of neuroscience studies so that it can epitomized in a discrete, brain-based response.

To Dussauge's analysis, we might add the insight that most studies assume sexuality to be static. That is, one's response to a Kinsey scale or similar question about sexuality is taken to be a stable characteristic. Such an approach does not allow for the fluidity of sexuality and sexual attraction. Despite the frequent use of pornography in such studies, as we will see below, they might at the same time be considered fairly conservative in their orientation because they privilege straightforward sexual acts and not the wide panoply of sexual

---

33. Dussauge, 131–32.
34. Dussauge, 144.

preferences that might be considered if one were to fully explore the topic from a diverse perspective.

To add to Dussauge's analysis, though, I'd like to focus on how neuroscientists could draw on research on genre and visual rhetoric as well as a feminist and queer studies lens. One large epistemological gap in these studies is their reliance on pornographic images to elicit sexual responses. These images are taken at face value, as natural stimuli for predictable sexualized responses. While neuroscience studies take for granted sexualized identities as pre-existing the images and the images simply producing inevitable responses, considered as genres, pornographic images can be understood as constituting participants based on conventionalized patterns of expectation and response. A rhetorical and cultural perspective suggests that these images are anything but natural, and neither are physical responses to them.

Here, I will examine two studies that typify the use of pornographic images as a stimulus. These studies were chosen in part because they are actually better than most: both seem to stem from a social perspective; one is even feminist. In other words, both studies sought to offer a somewhat richer interpretation of sexuality as culturally influenced. Yet, both studies end up naturalizing sexual responses by relying on the genre of the pornographic image without interrogating how those images work and what audiences they assume.

To begin, let's examine how one study deployed pornographic images to examine how women respond to visual sexual stimulation (VSS). In this study, published in the journal *Social, Cognitive, and Affective Neuroscience*, Charmaine Borg, Peter J. de Jong and Janniko R. Georgiadis begin with a promising acknowledgment that sex (especially VSS) is "under strong social control" and at least partially influenced by cultural norms.[35] This is an insight seldom acknowledged in neuroscience studies, which tend to sanitize stimuli (rhetorically and visually) in order to operationalize the concept in question. Additionally, this study differs from many in that it takes women as the primary study group rather than men.

Yet, the article quickly narrows the focus of the investigation to penetration. Since "the act of penetration lies at the very core of sexual activity," the authors reason, it should "carry considerable sexual incentive value" in a visual form.[36] For this reason, in the study, researchers showed "exclusively" or "predominantly" heterosexual women (all of whom had been in a heterosexual relationship for at least six months, were not virgins, and did not have

35. Borg, de Jong, and Georgiadis, "Subcortical BOLD Responses during Visual Sexual Stimulation Vary as a Function of Implicit Porn Associations in Women," 158.

36. Borg, de Jong, and Georgiadis, 159.

"sexual complaints") a series of pornographic images. These images, according to the authors, featured "hardcore coital interaction" with no faces and limited context. They featured only Caucasian, heterosexual couples, and included "easily recognizable features."[37]

Thus, in the experimental design, the authors are seeking to distill VSS into a common experimental stimulus that can then be universalized across study participants, regardless of sexual orientation, gender, race, or any other factor that informs sexual preferences. By not dwelling on this choice, and instead presenting it as an obvious, natural one, the authors rhetorically shift the reader's attention away from other options. Of course, the choice to feature penetration would seem like much less of a natural choice had the study design not already excluded homosexual and asexual women. Including a wider range of individuals would necessarily yield a wider range of sexual preferences and would likely call into question the centrality of penetration or its naturalness as the "core of sexual activity."

Further, the types of pornographic images available to researchers likely reflect conventions for that image genre. The authors do not describe exactly where they got their images, but there is certainly no shortage online for them to choose from. Mainstream pornographic images reflect a set of visual conventions; they are a genre. As Berkeley Kaite has argued, pornographic images contain "certain features—objects, discursive strategies, fetishized exchanges—[that] occur and recur with a vengeance and have a textual effect."[38] As genres, these images carry with them a set of formal conventions designed to elicit an embodied response. The pornographic image might be considered the paradigmatic form in this sense. If form "is the creation of an appetite in the mind of the auditor, and the adequate satisfying of that appetite,"[39] what genre does this better than porn? As Karen Ciclitira has noted, the types of pornographic images that are most widely available are premised on cultural topoi, or commonplaces, such as women's sexual willingness, men's sexual prowess, and specific types of bodies and body parts that widely circulate as representative of sexuality.[40] The authors do not describe the kinds of bodies depicted, but based on conventions for mainstream pornography we can assume that the bodies featured were of slender women and muscular men who represent conventionalized attractiveness. Exposure to these types of images likely conditions our responses to them. The fact that these images are carefully staged genres suggests that we should be careful not to interpret studies that

37. Borg, de Jong, and Georgiadis, 159.
38. Kaite, *Pornography and Difference*, viii.
39. Burke, *Counter-Statement*, 31.
40. Ciclitira, "Pornography, Women and Feminism," 285.

draw on them to mistake the stimulus (pornographic images) for the broader category of sexual desire as it occurs in other settings—that would lead us to neuroessentialism.

The researchers' choice of only Caucasian subjects for the stimuli is not explained in the article. The authors do not indicate the race of the study participants. We might assume, though, that the choice of only Caucasians in photos was because the study participants were Caucasian. Presumably, the authors assumed that showing pictures of people of another race might represent a confounding variable, or that race might disrupt the kind of response they sought to elicit. Perhaps they thought participants might not find photographic images of people from other races as sexually stimulating. Thus, the study presumes, but does not acknowledge, a rhetoric of interpellation and identification between study participants and the people represented in the images, wherein heterosexual participants are attracted to images of heterosexual participants who roughly resemble them in terms of sex and race.

Borg, De Jong, and Georgiadis were surprised by their findings: when shown images of hardcore sex (featuring penetration), participants exhibited neural responses that paralleled responses to the emotion of disgust as well as pleasure.[41] For the authors, this finding "raises doubt whether all brain activity induced by such stimuli can safely be assumed to be a signature of a positive sexual incentive value, which is nonetheless the dominant sentiment in VSS neuroimaging studies."[42] The authors seem unable to reconcile the idea that pleasure and disgust could commingle in responses to pornography. Yet a social and rhetorical perspective makes it much easier to understand why (if we take these findings to be true) pleasure and disgust might both characterize women's responses to pornographic images of penetration. For one, as Ciclitira explains in her study of women's attitudes toward and responses to pornography, feminist critiques of pornography have been influential. In interviews, Ciclitira found that some women drew on these critiques, noting that pornography was "degrading," that it involved abuse and exploitation of women, and so on. Others expressed ambivalence, noting that they found pornography arousing but felt guilty for feeling that way.[43] Another possibility, of course, is that the situation of the experiment itself evoked feelings of disgust. After all, the participants in question would be in an unfamiliar environment, with their heads and part of their bodies in an fMRI machine, knowing that researchers were recording their responses to the images they saw. How would one feel in that situation. Awkward? A bit creeped out? The point is that the

41. Borg, de Jong, and Georgiadis, 161.
42. Borg, de Jong, and Georgiadis, 165.
43. Ciclitira, 293.

situation itself is also rhetorical: it produces effects and cannot be excluded from the variables.

Thus, this example demonstrates that researchers seeking to engage cultural perspectives might do better if they seek out insights from researchers in cultural and rhetorical studies. In this particular case, cultural studies of visuality and rhetorical studies of science and genre can combine with psychological and neuroscience research to provide a richer interpretation of the study finding and a more nuanced understanding of what the experiment itself sought to measure (and how).

One might hope that a feminist orientation might lead researchers to develop a richer perspective. Yet a feminist orientation is not, in and of itself, sufficient: a feminist *rhetorical* perspective is needed. Mina Cikara, Jennifer L. Eberhardt, and Susan T. Fiske's article, "From Agents to Objects: Sexist Attitudes and Neural Responses to Sexualized Targets," offers a case in point. The authors are clearly operating from a feminist perspective. That is, they seek to answer a question about the effects of sexism, namely how the sexualization of women affects men. Sexualized images of women, they report, "disrupt[] the typical course of social cognition" by lessening the extent to which men viewing such images attribute mental states to the women pictured.[44] More specifically, they seek to show how men fail to grant agency to women pictured in sexualized poses or dress—especially in the case of men who hold sexist attitudes.

The study draws upon a model of sexism articulated by Fiske and her colleague, Peter Glick, one that is particularly savvy in that it identifies two valences to sexist beliefs. This model, the Ambivalent Sexism Inventory (ASI), suggests that sexism can fall along two intertwined tendencies: hostile sexism and benevolent sexism. Hostile sexism encompasses beliefs that express aggression toward women, such as the belief that women unfairly claim discrimination, that they seek to have power over men (not equality with men), and so forth. Benevolent sexism entails beliefs that express a more paternalistic attitude toward women, such as the belief that they need men's protection, that they are the more moral and sensitive of the two sexes, and so on. Both views, Fiske and Glick suggest, are harmful to women.[45] In the study by Cikara, Eberhardt, and Fiske, the ASI scale is used to find out whether men holding more hostile sexist beliefs are also more likely to view women as objects.

Clearly, there is much to appreciate about this study, one of few neuroscience inquiries to take an explicitly feminist orientation. Certainly, the effects

44. Cikara, Eberhardt, and Fiske, "From Agents to Objects," 541.
45. Glick and Fiske, "The Ambivalent Sexism Inventory."

of sexist beliefs are important, and studies like this one help us understand their deleterious effects. Yet, the feminist orientation of this study unfortunately leaves sex/gender and sexuality as unexamined essences (leading toward neuroessentialism) and fails to consider the rhetorical construction of the images used as stimuli (producing neurorealism).

A closer look at the study methodology demonstrates how this occurs. For one, the authors chose to include women in one part of the study, a test of verbal attribution of agency to images of men and women, but not in the second part, a brain imaging study using fMRI. Presumably, men are likely to see women as objects, but women are not likely to view other women that way. The study therefore assumes a male/female dichotomy in responses and overlooks the potential for a "brain mosaic." In other words, it would be interesting to ask whether women, as well as men, are affected by images that objectify other women—it seems plausible that this would be the case.

In the fMRI portion of the study, twenty-two men were recruited. While in an fMRI scanner, they viewed images of the following image types: "sexualized female," "clothed female," "sexualized male," and "clothed male." In these images, subjects were "smiling and gazing directly at the camera." The images were cropped from "mid-thigh to top of the head," and any clothing depicted was digitally altered to "minimize detail."[46] These images, we learn, were "of the sort that are frequently observed in public spaces (e.g., advertisements, billboards)."[47]

The sexualized images used in this study also constitute an image genre. Like the pornographic images used in the previous example, these images involve a set of recurring features and audience expectations. For one, in the US, at least, certain conventions govern what is displayed in these images and what is not. Typically, for instance, women are depicted in abbreviated clothing (bikinis, short shorts, cropped tops, etc.). Complete exposure of the body is not generally acceptable in public settings in the US; instead, "full frontal nudity" is reserved for X-rated spaces and media. Increasingly, though, these kinds of images remediate the techniques of pornographic images, using similar types of inviting gazes and suggestive poses.

In these studies, then, assumptions about sexuality, sex, and gender are baked into the study design, from prescreening forms that require participants to identify their sex, gender, and/or sexual preference to the images used to invoke sexuality itself.

---

46. Cikara, Eberhardt, and Fiske, 543.
47. Cikara, Eberhardt, and Fiske, 542.

A NEURORHETORICAL APPROACH TO
SEX, GENDER, AND SEXUALITY

Despite some promising directions, the studies that I examined above could benefit from raveling together humanistic perspectives. To start with, none of these studies takes an intersectional approach. As Jacqueline Jones Royster and Gesa Kirsch describe it, an intersectional feminist approach moves us "beyond establishing the existence of basic issues of identity and representation toward ever more complex analyses of the social, geopolitical, and cultural dimensions of analytical paradigms as they are infused with complicated relationships and interrelationships in light of various contexts, conditions, and processes."[48] The term *intersectionality* was coined by legal scholar Kimberlé Crenshaw, who describes it as a way of understanding how "race and gender intersect in shaping structural, political, and representational aspects" of inequalities black women face.[49] Since then, the term has been used to describe a method of analysis that considers not just race and gender but also disability, ethnicity, class, nationality, sexual orientation, and so on as intersecting factors shaping individuals' experiences.

Intersectional methods emphasize complexity, but neuroscience research (at least as it is currently practiced) tends to require researchers to focus on one set of variables at a time and to suggest whether a correlation does or does not exist between those variables. Barbara Tomlinson notes that, for this reason, "positivist disciplinary research methods . . . may be at odds with and undermine intersectionality's ability to provide radical critiques of power."[50] Despite this caveat, it seems important to at least try to articulate how an intersectional approach might inform neuroscience studies of sex and sexuality. Rita Dhamoon usefully outlines how this approach might be undertaken in her home discipline, political science, but she reflects on its usefulness for social science research in general. This approach, she argues, "foregrounds a richer ontology than approaches that attempt to reduce people to one category at a time, it treats social positions as relational, and it makes visible the multiple positioning that constitutes everyday life and the power relations that are central to it."[51] Here, I draw on Dhamoon's framework to suggest some of the challenges of an intersectional approach for social and cognitive neuroscience. In particular, working within the rhetorical-material infrastructure of neuro-

---

48. Royster and Kirsch, *Feminist Rhetorical Practices,* 48.
49. Crenshaw, "Mapping the Margins," 1244.
50. Tomlinson, "To Tell the Truth and Not Get Trapped," 1002.
51. Dhamoon, "Considerations on Mainstreaming Intersectionality," 230.

science—especially the dictates of positivism—creates unique challenges for neuroscience that will require creative solutions.

Dhamoon sketches out four levels of analysis for intersectional approaches: identities, categories of difference, processes of differentiation, and systems of domination. At the level of identity, social science researchers focus on how people understand themselves and their experiences in light of social identities, such as "Black woman" or "gay man." In neuroscience, research that focuses on identities often takes up a related strategy, seeking to identify differences between individuals of one identity group or another. A study might compare men and women but not consider racial identities of those groups. Sometimes they are not even reported, so it is hard to know whether groups of participants reflect a racially diverse pool or not. Further, given the legacy of evolutionary psychology and positivism, often these identities are taken as natural or biological. Thus, a study that reveals differences between men and women can be interpreted as supporting an evolutionary understanding of sex differences because it glosses over other factors that might indicate differences that are likely to be culturally conditioned. An intersectional approach would point out that any identity chosen as a dimension of analysis is a cultural location infused with sociocultural and rhetorical factors. What it means to be a woman, for instance, changes dramatically across time, space, culture, and so on. As Dhamoon puts it, an intersectional approach "exposes the myth that identities naturally pre-exist and the fallacy that subjects *have* [rather than perform] identities."[52] Yet, in research studies those categories are taken as single and universal, an approach that easily leads to essentialism.

Focusing on intersectional identities would challenge researchers who typically focus on a single dimension of analysis. Methodologically, a single variable makes analysis easier from start to finish. One approach for researchers might be to take a more fine-grained approach. Even in a traditional study of male versus female subjects, researchers might gather demographic data that would indicate the composition of the participant pool in terms of race, class, sexual orientation, or any other quality that might be relevant to the phenomenon at hand. Alternatively, researchers might seek more specific groups of participants (Black women, say, as opposed to "women"). Nonetheless, this approach still risks essentialism: the whole point of intersectionality is that multiple variables make up identity in ways that may be highly individualized once multiple factors are taken into account. Determining which variables are relevant in which cases becomes part of the challenge.

---

52. Dhamoon, 235.

Another option might be to study, rather than sex or gender differences, the process of categorization itself. What Dhamoon terms "categories of difference" might refer to how people categorize other individuals as members of a given group. The studies of voice (Tesink et al.) and gait (Jordan, Fallah, and Stoner) work along these lines, since each sought to identify how people group other people into gendered categories. To some extent, these studies lean toward an intersectional analysis. For instance, Tesink et al. included age, sex, and social class as dimensions of analysis, while Jordan, Fallah, and Stoner included gender-ambiguous as well as male and female examples of gait. We might imagine how these studies could be extended to include a wider range of variables, including, for instance, race and ethnicity, both of which might enter into analysis of voice or gait. Our voices reflect not only our gender and class but also the cultural and linguistic communities we belong to, and these are likely to intersect with race and ethnicity. Is a woman speaking with a particular accent viewed as more feminine or less? Does it depend on the accent? Similarly, the way we walk (which may seem like a more universal factor) may reflect anything from the kinds of physical activities one enjoys to preferred style of shoe to the typical speed at which people walk in your community or city. Once you factor in those elements, it becomes apparent that gender would intersect with various cultural factors that could be linked to social class, ethnicity, race, and so on.

Another promising approach would be to study the processes of gendering, racialization, and other social developments that constitute identities. Dhamoon defines the focus on processes as describing how "subjectivities and social differences are produced, such as through discourses and practices of gendering, racialization, ethnicization, culturalization, sexualization, and so on." These studies also do not get at how it is that certain behaviors come to be associated with a specific gender (race, class, etc.) in the first place, or whether new associations can be induced experimentally.

To address these questions, researchers might study the effects of systems of differentiation. Here, Dhamoon refers to "historically constituted structures of domination such as racism, colonialism, patriarchy, sexism, capitalism, and so on."[53] Indeed, research suggests that even implicit stereotypes (which are presumably earned early in life) are malleable and can be influenced experimentally using priming. To date, however, research in this vein tends to focus on a single variable and often avoids questions of how, for instance, racist and sexist attitudes interact. In one study, Laurie Rudman and Julie Phelan showed female students images of men and women in traditional and nontraditional

---

53. Dhamoon, 234.

occupations. Their study found that, when primed with stereotyped gender roles, women showed increased automatic gender stereotypes in an Implicit Association Test. They also found that, when primed with nonstereotyped gender roles, participants actually showed reduced interest in male occupations and lower self-perceptions of leadership ability, a finding they interpret as suggestive of the potential for images of strong, successful women to have a backlash or intimidating effect. While intriguing, the study included a sample that was 89 percent white.[54] The results would be better interpreted as indicative of *white* women's responses to gendered stereotypes. Since the study was conducted in a university setting (presumably Rutgers, where the researchers work), we can also assume that the participants were largely from middle- or upper-middle-class backgrounds. Whether and how women from other backgrounds would respond to such a test remains an open question.

In addition to drawing on intersectional theories, studies of sex, gender, and sexuality might benefit from humanistic theories of vision and visual culture. Neuroscientific studies of sexuality, especially, rely frequently on visual images that partake in a set of cultural practices, or a visual culture, that shape how those images are viewed and what they are taken to represent. By raveling together neuroscience approaches with those drawn from humanistic fields, we can generate a deeper understanding of *looking* as a cultural practice. In *Techniques of the Observer,* Jonathan Crary explains that "vision and its effects are always inseparable from the possibilities of an observing subject who is both the historical product *and* the site of certain practices, techniques, institutions, and procedures of subjectification."[55] An observer is "one who sees within a prescribed set of possibilities, one who is embedded in a system of conventions and limitations."[56] These conventions include image genres.

David Park's study of Civil War image genres demonstrates how different types of images included in Civil War–era newspapers in the US required conventionalized ways of seeing. Park notes that images featuring battle scenes relied on realist ways of looking inculcated by landscape painting and photography, whereas portraits of famous figures inculcated an epideictic orientation to the subject.[57] For Park, each new image genre (used in Civil War newspapers) "was based on its own set of assumptions regarding its relationship to the outside world, possessed its own rhetorical sensibility, and generated its

54. Rudman and Phelan, "The Effect of Priming Gender Roles on Women's Implicit Gender Beliefs and Career Aspirations," 195.

55. Crary, *Techniques of the Observer,* 5.

56. Crary, 6.

57. Park, "Picturing the War," 291, 299.

own range of national symbols."[58] These images, in turn, influenced how actors understood the Civil War, helping to constitute viewers' identities as well as their attitudes toward the subjects.

As Cara Finnegan has explained, one practice shaping photographic images, in particular, is the naturalistic enthymeme: we tend to take images as "true" or "real" representations.[59] The photograph embeds the visual convention of a "fixed, monocular eye that [bears] an apparently identical resemblance to nature"—despite the fact that the photograph differs from human vision in several ways.[60] For instance, we do not perceive the world as bounded by a rectangular frame, as in a photograph; nor do we see objects as uniformly focused across our field of vision (instead, humans only see objects sharply at the center of our field of vision). The photographic image is accepted as "real" to us, but it actually tropes and exaggerates reality: the photograph is meant to stand in for the real thing, yet it augments reality by being sharper, more colorful; it is hyperreal. As Jean Baudrillard explains, the hyperreal emerges when images no longer refer to an origin or reality, nor do they even seek to dissimulate a reality; instead, they are used to "dissimulate the fact that there is nothing behind them." The images have no "relation to any reality whatsoever."[61] One of Baudrillard's examples is Disneyland, a "play of illusions and phantasms";[62] pornography operates similarly.

For contemporary pornographic images, produced with the help of digital technologies such as airbrushing and computer animation, it is not simply that an image of sex stands in for the reality of sex; the image masks the fact that the sexual acts themselves are fake, an elaborate masquerade based on signs (a facial expression, an item of clothing, a prop) and often bodies that have themselves been reconfigured through plastic surgery to become signs.[63] The bodies are often equally exaggerated: skin may be waxed, tanned, and oiled; body parts may be surgically enhanced; camera angles emphasize curves and shapes. The viewer of the pornographic image is expected to suspend disbelief, taking what is depicted as a spontaneous, real sexual act, despite knowing that the image displays models or actors in an exaggerated sexual choreography.

------

58. Park, 290.

59. Finnegan, "The Naturalistic Enthymeme and Visual Argument," 135.

60. Finnegan, 141.

61. Baudrillard, "Simulacra and Simulations," 172–73.

62. Baudrillard, 174.

63. The use of the term *porn* to describe other kinds of images that circulate through contemporary media, such as food, yarn (in knitting and crocheting circles), and the like attests to the extent to which photographic and digital technologies honed in porn have spread to other settings and, along with them, certain ways of looking and certain kinds of hyperrealities.

Yet viewers continue to view pornographic images as erotic, and are conditioned to respond in certain ways, because pornographic images constitute an image genre or a rhetorical form. In Burke's words, rhetorical form itself constitutes a rhetorical appeal in that it "gratifies the needs which it creates"; in the case of pornographic images, we might find an especially apt example of this.[64] That is, viewers of a pornographic image expect to have a specific embodied response (sexual arousal) and to respond in certain ways (perhaps not just viewing the image but acting on their feelings of arousal).

These images also constitute sexed and gendered identities for those who produce, view, and are depicted in them. Anis Bawarshi argues that "genres have this generative power because they carry with them social motives—socially sanctioned ways of 'appropriately' recognizing and behaving within certain situations—that we as social actors internalize as intentions and then enact rhetorically as social practices."[65] A straight man, for instance, is socially conditioned to respond sexually to pornographic images that prominently feature women (either with other women or with men) but is expected to be turned off by pornographic images featuring gay male sex. Women, in contrast, are often expected to be turned off by all pornography, although this assumption seems to be changing as pornography becomes more and more mainstream.

Images, especially images of people, constitute a rhetoric of interpellation. As Maurice Charland explains, "interpellation occurs at the very moment one enters into a rhetorical situation, that is, as soon as an individual recognizes and acknowledges being addressed."[66] Image genres featuring people, especially, elicit performances of social identities, including ones that are sexualized, gendered, raced, and classed. When viewing images of people, identities are instantly called into play through processes of identification and disidentification: the person pictured is either "like me" or "not like me"; "attractive to me" or "not attractive to me." The image therefore calls into play, or constitutes, gender, race, and sexuality. The pornographic image does so especially powerfully because it is designed to elicit a visceral, embodied response.

Experiments in neuroscience that rely on images similarly depend upon image genres and their attendant ways of seeing and responding. Even studies that take a relatively sophisticated view of gender and culture still tend to naturalize images and fail to take into account the rhetorically and culturally inscribed ways of seeing that inform them. Understanding these stimuli as

---

64. Burke, *Counter-Statement*, 138.
65. Bawarshi, "The Genre Function," 341.
66. Charland, "Constitutive Rhetoric," 138.

image genres offers a richer way of interpreting the findings as well as possibilities for richer study designs.

Of course, one might question the very idea that there is one "core" of sexual activity that would be the best type of image to show women in order to elicit sexual responses. As Marti Hohmann puts it, "there is no 'natural' or 'good' sexuality to be recovered from pornography representations";[67] a focus on penetration reflects a goal-oriented model of sex in which orgasm (and especially male orgasm) is the end point, not a natural "core" of sexual activity. Research on men's and women's responses to pornography suggests an interesting difference: "heterosexual men show greater responses to depictions of women than to depictions of men, whereas gay men show the converse pattern. In contrast, both heterosexual and lesbian women respond substantially to both depictions of women and depictions of men."[68] In other words, the choice to depict heterosexual, penetrative sex does not necessarily reflect women's preferences. In one study, Meredith L. Chivers, Michael C. Seto, and Ray Blanchard used a wide range of visual stimuli, including pictures of naked people, depictions of people masturbating, depictions of oral sex, and even pictures of bonobos having sex. For each of the image types involving sexual activity, the researchers included male-male, female-female, and male-female examples. Here, Chivers et al. found that women responded to sexual activity regardless of whether the people pictured in it reflect their own sexual preferences—or even whether they were depictions of the same species.[69]

In addition, the study designs described here assume that sexual orientation and gender are both static essences. Participants were defined based on their declared gender and sexual orientation, which precludes the possibility that both gender and sexuality might be something more than a single, static entity. L. Ayu Saraswati argues that we might better understand sexuality as a fluid process, taking on various forms in different situations and across time. She sees sexuality as both something constituted by the interpellations of others, and something that we perform for an audience: sexuality "necessitates an audience (the other) in order for it to be fully articulated."[70] We might apply this concept to gender as well, understanding it not as an a priori identification or essence but as something that gets invoked within a rhetorical situation that includes an audience.

All of this is not to say that neuroscience studies cannot tell us something interesting about sex, gender, or its intersections with other social differences.

67. Hohmann, "The Politics of the G-Spot," 23.
68. Chivers, Seto, and Blanchard, "Gender and Sexual Orientation," 1108.
69. Chivers, Seto, and Blanchard, 1116.
70. Saraswati, "Wikisexuality," 594.

If anything, these studies suggest that sex and gender are indeed rhetorical in the deepest sense of the word: they are both the result of embodied effects and productive of those effects (both differences and similarities). But to take those effects as simply natural and inevitable overlooks the ways that those effects are also embedded in cultural and rhetorical modes of production. The next chapter shows how a similar tendency shapes neuroscience research into politics.

CHAPTER 5

Neuropolitics

IN 2015, as the US presidential election season swung into high gear, candidates made immigration a central issue in both the Republican and Democratic primaries and in the eventual presidential race. On the Republican side, iconoclastic presidential candidate Donald Trump rallied supporters by espousing extreme views on immigration. He famously proposed a ban on Muslims immigrating to the US,[1] the construction of a wall across the US–Mexico border,[2] stricter enforcement of immigration laws by tripling the number of Immigration and Customs Enforcement officers, and an end to birthright citizenship to discourage illegal immigration.[3] As Trump's competitors sparred over who was taking the toughest or most pragmatic angle on immigration, voters picked up on the underlying message. One voter quoted in a *Wall Street Journal* article admitted that she was "fearful of letting Muslims in. When I've gone to other countries they called me an infidel; I don't want to be an infidel in my own country, and it's a little scary."[4] Another, com-

---

1. Diamond, "Donald Trump: Ban All Muslim Travel to U.S."
2. Weissmann, "Donald Trump Explains His Ridiculous Plan to Make Mexico Pay for a Border Fence."
3. Donald J. Trump for President, "Immigration Reform." Information removed from website in 2017; version retrieved from archive.org.
4. Hart, "In 2016 Voter's Anxiety and Fear, a Message to the Establishment."

menting on a blog post on the far-right website Breitbart.com, went a bit
further:

> What sane person would want Third World Jihadists moving in next door?
> We are at war; under attack by radical Islam. Barack Obama and Hillary
> Clinton are trying to take our guns away? They are pushing to import MORE
> terrorists? This is insanity . . . lunacy. Anyone that supports the notion that
> ILLEGAL immigration (be it border crossers or Obama's terrorists via fiat)
> is an American value is un-American. Promoting the destruction of law and
> order in your own country is treasonous. Disguising it in some notion of
> humanitarianism is nothing but deception.[5]

Other commenters echoed the incredulity that anyone—namely Democrats—
could support bringing immigrants into the country. And then a commenter
named BaseballSince10 wrote this comment: "I'd like to point out (sorry I
can't remember the sources) that scientific studies of the brains of Lefties and
Righties point out various brain differences, as if there is a certain amount of
pre-disposition to be either hard-wired one way or the other."[6] This struck a
chord with another respondent, who noted, "when it comes to progressives
. . . the intellectual dishonesty and psychological contortions they use to justify their twisted anti-logic . . . is seemingly beyond all rational explanation."[7]
Unable to fathom how someone could hold a different point of view, these
commenters looked to neurological differences for an explanation.

Similar arguments have been offered to explain why "fake news" websites
hold sway over readers.[8] An article in *The Week* that appeared shortly after
the 2016 presidential election compared the glut of fake news to the state-run
media machine in Russia and quoted Vitaliy Katsenelson, a Russian investor:
"'Russia's propaganda works by forcing your right brain (the emotional one)
to overpower your left brain (the logical one), while clogging all your logical
filters.'" How else to explain why readers unquestioningly accept and share
propaganda that is so outlandish?

In a political climate that seems increasingly hyperbolic and polarized, in
which those on the right and the left just can't seem to agree on anything, neuroscientific explanations can seem appealing—something both parties might
agree on. Perhaps, as Betty Vine argues on the website *BrainWorld*, those

---

5. May, "Poll: Majority of Voters Fear 'Homegrown Jihadists,' Most Oppose Accepting
Syrian Refugees."

6. May.

7. May.

8. Lange, "American *Pravda*."

on the right are neurologically primed to respond to fear—and that might explain why they respond so strongly to the issue of immigration. Maybe liberals really are better at weighing conflicting information, making them more comfortable with uncertainty and less susceptible to fear.[9] Yet, a closer rhetorical analysis of these brain-based explanations for political behavior would lead us to question whether these differences are simply natural and hardwired. Instead, a rhetorical perspective would ask us to consider how social and cultural practices may in fact materialize neurologically (or how they might be mutually constitutive) and how neuroscience studies themselves draw upon a rhetorical infrastructure based on a Western political model and then reify that model as natural or innate.

To demonstrate how social and cognitive neuroscience studies approach politics, I will first consider how neuroscientists typically study politics. I focus here on core scientific studies that are commonly cited as evidence of neurological differences between liberals and conservatives. These articles assume that political differences are stable, both in terms of their embeddedness in brain structures and in terms of their universality across time and place. In particular, they overlook the significant role that racial beliefs and attitudes play in American politics. Differences among racial groups in political orientation call into question the idea that liberalism or conservatism are simply neurologically or genetically based; to take that view promotes a dangerously solipsistic and whitewashed understanding of politics. In the second part of the chapter, I argue that a rhetorically based understanding of political discourse, one that considers the circulation of discursive and affective political rhetorics, can help to deepen a neuroscientific understanding of politics.

HOW NEUROSCIENCE STUDIES POLITICS

Let us first ravel back, considering the fundamental metaphors and guiding definitions that shape inquiry into the neuroscience of politics. Neuroscience studies of politics often begin by setting out an antithesis (or sometimes a double antithesis) between liberals and conservatives that depends on synecdoche. Just as "male" and "female" brains stand in for men and women in the case of sex/gender research, "right" and "left" brains stand in for conservative and liberal individuals (in the US context, Republicans and Democrats). The antithesis of right/left is less a metaphor than an antithesis that relies on synecdoche.

9. Vine, "The Neuroscience of Politics."

The titles of articles in this line of research evoke this antithetical logic. The first article to be considered here, published in 2007 by David M. Amodio, John T. Jost, Sarah L. Master, and Cindy M. Yee, appeared as a brief communication in *Nature Neuroscience* in 2007. Evoking neurorealism and neuroessentialism, the title itself, "Neurocognitive Correlates of Liberalism and Conservatism," constitutes both liberalism and conservatism as concrete, oppositional qualities that can be identified in the brain. The second article, "The Political Left Rolls with the Good and the Political Right Confronts the Bad: Connecting Physiology and Cognition to Preferences," appeared in the *Philosophical Transactions of the Royal Society* in 2012. Written by Michael D. Dodd et al., its title evokes a double antithesis between left/right and good/bad. The third article similarly dichotomizes political orientations. Published by Darren Schreiber et al. in *PLoS ONE*, "Red Brain, Blue Brain: Evaluative Processes Differ in Democrats and Republicans," clearly sets up political orientations as types of brains—a clear example of neurorealism. Like the Dodd et al. article, it uses a double antithesis, equating red brains with Republicans and blue brains with Democrats.

In fact, "right" and "left" are not neutral descriptors but metaphors commonly used to organize and group political orientations. The categories "right" and "left," as Chantal Mouffe would argue, reflect not simply a natural neurological state but the history of liberalism itself. The right–left spectrum is itself a cultural construct, one with a relatively recent history dating back to the time of the French Revolution (when supporters of the crown and radicals literally sat to the right or left of the king). This metaphor, as cultural critic Susan Sontag explains, "distributes, and polarizes, attitudes and social movements" according to a deeper, older logic that constitutes politics using metaphors of the body, whose left and right sides are united only by the "head of state," he or she who provides discipline and rule. Comparing the state to a body in this way, Sontag explains, "makes an authoritarian ordering of society seem inevitable, immutable."[10] Today, however, these metaphors have become naturalized in public discourse to the extent that we no longer consider them metaphors. As a result, they seem like a natural, obvious way of describing political orientations for these researchers.

Researchers then argue these oppositions into place in the introduction to each article by citing previous psychological research. For instance, Amodio et al. begin with this sentence: "Political scientists and psychologists have long noted differences in the cognitive and motivational profiles of liberals

---

10. Sontag, *Illness as Metaphor*, 94.

and conservatives in the USA and elsewhere."[11] Similarly, in their study, Schreiber et al. start with a similar summary of current research: "A large body of research suggests that liberals and conservatives differ on important psychological characteristics."[12] By generalizing based on past research, these researchers establish political differences as not only readily observable in the outside world but also as supported by scholarship.

Dodd et al. take a more philosophical tack in their introduction, observing that these antithetical relations are a universally troublesome feature of human societies: "The most intense cultural conflicts tend to be disputes over the proper way to structure and maintain mass-scale social life. Accordingly, whether within or across national boundaries, disagreements regarding politics . . . are much more likely to lead to acrimony and even violence than, say, disagreements over preferred personality traits or taste in arts. Politics can affect the lives of others in a way that personality and taste do not. It has been a flashpoint over the centuries and serious attempts at understanding conflict must address the reasons for political differences."[13] The authors seem to be imagining a world in which political differences might be remediated, in which conflict might somehow not be constitutive of politics and of human society in general.[14] The introductions to these articles, then, establish a clear binary (left/right, liberal/conservative, we/they) that is purportedly observable in everyday life. They present this binary as natural, and not itself subject to inquiry—that is, researchers do not ask why that difference may exist and how it is produced. Neuroscientists then assume that political differences reflect biological (neurological) differences.

Based on these oppositions, researchers then structure a cognitive profile for each group, although that profile differs from study to study. "Liberalism" and "conservatism" (or "Left" and "Right") are enacted slightly differently in each experiment. In some cases, two elements that are not truly opposites are made into opposites by virtue of being aligned with the opposite terms, Left and Right. Antithesis is a rhetorical figure that, in Fahnestock's terms, "has often been used in scientific arguments to construct terms into opposi-

---

11. Amodio et al., "Neurocognitive Correlates of Liberalism and Conservatism," 1246.

12. Schreiber et al., "Red Brain, Blue Brain," 1.

13. Dodd et al., "The Political Left Rolls with the Good and the Political Right Confronts the Bad," 640.

14. Such a view is idealistic, but not too realistic. Political theorists view conflict as central to human society, even beneficial, so long as it can ultimately lead to productive outcomes. For instance, Chantal Mouffe would argue that antagonism (a complete we/they opposition between enemies) is destructive, but that agon (the oppositional relationship of adversaries) is productive and in fact necessary for a democratic society to flourish. See Mouffe, *On the Political*, 20.

tion that are not forceful opposites to begin with."[15] One way to do this is to use a double antithesis, in which a set of clearly opposed terms maps onto a set of terms that may not so clearly be opposed, but that can be passed off as opposed by being aligned with the original term. In Amodio et al., for instance, the authors report that conservatives "have been found to be more structured and persistent in their judgments and approaches to decision-making," while liberals show greater tolerance for ambiguity, perplexity, and new experiences.[16] In this way Amodio et al. map out a double antithesis that pushes apart two groups of people (liberals and conservatives) while simultaneously working to establish the secondary antithesis as antithetical in the first place. Here I mean that structured, persistent decision-making's true opposite would actually be completely impulsive decision-making, not tolerance for ambiguity or complexity per se. By the same token, the true opposite of tolerance for ambiguity would be a preference for certainty. Nonetheless, the rhetorical logic underpinning the opposition between liberals and conservatives pushes us to accept the secondary premise as valid.

We see a similar double antithesis working in Dodd et al. Here, the authors first establish that individuals may be driven more by the desire to avoid aversive stimuli (fear, disgust) or by the desire to seek out hedonistic stimuli (pleasure, enjoyment) and then correlate those two tendencies with "right-of-centre" and "left-of-centre" political orientations, respectively. The double antithesis here separates out physiological responses that all humans share, forcing them apart and aligning them with political orientations that are likewise polarized. The authors then map these oppositions onto specific political beliefs: "From this perspective, it makes sense that people who are more attentive and responsive to hedonistic stimuli would support tax dollars being spent on the arts and national parks, just as it makes sense that people who would be more attentive and responsive to aversive stimuli would advocate policies promoting moral purity and harsh treatment for norm violators."[17] Here again, the logic of these passages depends upon the double antithesis, which creates binaries, forces them apart, and then articulates them with specific topics (funding for the arts, moral purity). Usually, pleasure is opposed to pain, not disgust, but in this case they are rendered as opposites because they are aligned with a pair of terms more readily seen as opposites, Right and Left.

In Schreiber et al., the authors ascribe "stronger attitudinal reactions to situations of threat and conflict" to conservatives and a tendency to "seek out

---

15. Fahnestock, *Rhetorical Figures in Science*, 70.
16. Amodio et al., 1246.
17. Dodd et al., 641.

novelty and uncertainty" to liberals.[18] This understanding is similar to that set out by Dodd et al.; here, though, the aversive stimuli associated with conservatives is tagged as "threat" as opposed to "disgust," and the hedonistic stimuli associated with liberals is "novelty" as opposed to pleasure. The rhetorical figure (double antithesis) is the same, and it works to make whatever differences are introduced seem like they are natural and commonsensical.

To study politics within this neuroscientific paradigm, moreover, researchers rely on neurorealism: they take for granted that political orientations exist, that they are identifiable, and that they have a neurological basis. For instance, Amodio et al. argue that political orientation must have a neurological basis because it has been shown to "to be heritable, evident in early childhood, and relatively stable across the lifespan." For this reason, they continue, a "basic neurocognitive mechanism" must be involved.[19] They then locate this mechanism in a specific region of the brain, the Anterior Cingulate Cortex. While Amodio et al. refer to a "mechanism," other researchers might refer to political orientation as embedded in "networks," as Zamboni et al. do in an article where simultaneously they seek to identify "how the psychological architecture of political beliefs is organized and where it is represented in the brain."[20] These metaphors of "mechanism" or "architecture" constitute neurorealism, which is considered problematic in popular accounts, but here it appears in a scientific article. Appealing to the heritability or "architecture" of political beliefs renders them concrete, real, and therefore operationalizable in an fMRI study.

Now, let's ravel out how neuroscientists tend to study politics. The next step is to select participants who will reliably display the characteristics assigned to each political orientation. The measures used to select participants help to constitute the political orientation; they should be recognized as "actants," to use Latour's term, that materially shape that which is being measured or, in Barad's terms, as "relevant features" of the experiment that in fact constitute what is being observed.[21] Despite the many ambiguities we might attach to them, all these measures are used to slot participants into a clear right or left category, eliminating the nuances that we might otherwise recognize in political orientations.

To identify participants' political orientations, researchers sometimes simply recruit participants who have previously self-identified according to politi-

18. Schreiber et al., 1.

19. Amodio et al., 1246.

20. Zamboni et al., "Individualism, Conservatism, and Radicalism as Criteria for Processing Political Beliefs," 368.

21. Barad, Meeting, 106.

cal orientation. In the Schreiber et al. study, participants were recruited in this way, matching participants recruited for a previous study to a list of political party registration records.[22] This approach seems straightforward, but it has some significant limitations. For one, not all eligible voters register with a political party; the largest group of voters in the US are those who are registered as independent—in 2016, this represented 34 percent of voters.[23] Further, the two-party system in the US often creates a forced choice between Republican, Democrat, and independent. As of 2016 only twenty-seven states allowed voters to register as Libertarian, the third-largest party in the US. In fact, an additional thirty-three political parties exist in the US but do not show up in studies that simply select from Republican- or Democrat-identified voters. These include a range of socialist-identified parties, parties formed around single issues (legalization of marijuana, copyright issues), parties linked with particular beliefs (from white-supremacist to religious values of various kinds), and parties aligned with specific identities or subject-positions (veterans, Black Panthers). By excluding participants who identify with these political parties, neuroscience researchers are in effect purifying or reducing what is, in Burke's terms, a veritable "Human Barnyard."[24] The problem of the "excluded middle" characterizes many neuroscience studies of political orientation.

A second option involves asking participants to self-report on a scale, as in Amodio et al., where the authors asked participants to simply choose an answer on a single-scale measure of political attitudes (from extremely liberal to extremely conservative). While a scale might seem like a neutral way of measuring political orientations, we must recognize it as in fact constituting those orientations, insofar as they must first stabilize and then essentialize what are, in practice, messy collections of beliefs, values, opinions, and habits. If given the option to elaborate, many people would likely clarify their stances: "I'm fiscally conservative but socially liberal" or "I'm conservative on abortion but liberal on immigration" or what have you. Note, for one, that here again alternative political views are excluded from this scale, including Marxist, Libertarian, Green, and so on. These political orientations map uneasily onto a single scale, and would require a more complex, multilayered way of measuring political orientations. Scales, then, require participants to perform synecdochally, choosing a part of their political beliefs to represent the whole.

---

22. Schreiber et al., 1.
23. Pew Research Center, "The Parties on the Eve of the 2016 Election."
24. Burke, *Rhetoric*, 23.

In practice, researchers sometimes *only* select participants who identify as extremely liberal or extremely conservative, further sanitizing their sample. Dodd et al., for instance selected their participants from a group of forty-eight individuals who had already responded to a questionnaire. In the article, the authors note that they selected those forty-eight because "they were the individuals most clearly falling on either the political left or the political right."[25] In this way, the researchers in effect "salted the mine," choosing participants who they thought would produce clearly demarcated results.

The third type of approach used to determine political orientations seems, on the face of it, more nuanced than the first two. In this approach, participants respond to a psychometric test, the Wilson-Patterson scale, in which they indicate their agreement with short phrases or single words, such as "death penalty," "gay rights," "capitalism," or "unions." In the Dodd et al. study, participants responded to the first type of question (choosing how they identified on a scale from liberal to conservative and on a scale from Democrat to Republican), but then participants completed a version of the Wilson-Patterson scale. These scales work by multiplying a single scale (liberal to conservative) out into multiple scales. Yet, the same challenges persist.

The Wilson-Patterson scale was introduced in 1968 by Glenn D. Wilson and John R. Patterson. Claiming that existing measures at the time were susceptible to "agreement response bias" (the tendency for participants to agree with prompts for reasons other than their actual identification with those ideas, such as the eloquence of a statement), Wilson and Patterson offered a new measure, one that avoided the lengthy descriptions of opinions that could potentially lead respondents to answer on the basis of the eloquence of the opinion rather than its content. Wilson and Patterson recognized that the form of the questionnaire was itself persuasive: participants could be led to agree with a statement of a political opinion on the basis of its style. Searching for a more transparent (hence, presumably arhetorical) means of presenting political ideas, Wilson and Patterson decided to simply present readers with "brief labels or catch-phrases" that represented common issues.[26] They felt that these short phrases would better capture participants' initial, affective response to an issue, which they believed better predicted political behavior.[27] Presumably, this approach avoided the problem of rhetorical style, since sentences were reduced to phrases.

25. Dodd et al., 642.

26. Wilson and Patterson, "A New Measure of Conservatism," 265.

27. Wilson and Patterson.

This list of catchphrases that Wilson and Patterson included began with a set of assumptions about what conservatism is, which they listed as follows:

- religious fundamentalism
- right-wing political orientation
- insistence on strict rules and punishments
- intolerance of minority groups
- preference for conventional art, clothing, and institutions
- anti-hedonistic outlook
- superstitious resistance to science

To test those core beliefs, they came up with a list of fifty themes, such as "patriotism," "socialism," "legalized abortion," and "death penalty."

By selecting participants that strongly respond to these stimuli, researchers try to ensure that they will have "statistically significant" results. That is, including a wider range of political moderates would not produce a stark antithesis between right and left, and it could not lead to strong statements about neuropolitical differences. For instance, Amodio et al. conclude from their study that, "although a liberal orientation was associated with better performance on the response-inhibition task examined here, conservatives would presumably perform better on tasks in which a more fixed response style is optimal."[28] We see a similar antithesis in Schreiber et al., who conclude that "Republicans more strongly activate their right amygdala, associated with orienting attention to external cues," while "Democrats have higher activity in their left posterior insula, associated with perceptions of internal physiological states."[29] These brain regions and attendant behaviors are not necessarily opposed, but the logic of these two sentences puts them in an oppositional arrangement because Republican/Democrat has already been established by the authors as an antithesis and baked into the experimental design.

In sum, we see neurorealism and neuroessentialism at work in the rhetorical structure of this experiment, not just in the article itself. We might say that these stasis points are embedded in the material practices of neuroscience, including the psychometric scales used to measure political values and beliefs. Thus, neuroscientific studies tend to view political orientations as fairly static, binary, and universal. From a rhetorical perspective, though, we might see political orientations as assemblages of discourses, practices, and affects that circulate and change over time and place. This viewpoint fosters a differ-

---

28. Amodio et al., 1247.
29. Schreiber et al., 1.

ent approach to neuropolitics; in the next section I will show how raveling together neuroscientific and rhetorical theories can lead to such an approach.

A NEURORHETORICAL APPROACH TO POLITICS

To develop a neurorhetorical approach, we might draw on research in political theory and rhetoric that deepens our understanding of politics and its centrality to human life. Researchers in the neurosciences posit that political differences are essential to human cognition. Researchers in humanistic fields might agree about the centrality of political differences, but theorize them differently. Carl Schmitt, whom Mouffe draws on extensively, argues that politics arises out of a friend/enemy or we/them distinction.[30] This agonistic relationship is, in fact, the basis for rhetoric.

Yet rhetorical theory would not map this opposition onto neat, mutually exclusive divisions of "right" and "left." In *Bodily Arts*, Debra Hawhee argues that early sophistic rhetoricians thought of *agon* as "a gathering of forces—cultural, bodily, and discursive."[31] In an agonistic situation, there is an opposition or tension between these different forces—but the opposition is not necessarily clear-cut. As Burke explains in *A Rhetoric of Motives*, the rhetorical situations are not possible in cases of "pure identification" or in cases of "absolute separateness"; only when there is some kind of "mediatory ground [is] their communication [made] possible."[32] This mediatory ground, or what Hawhee calls "a gathering of forces," occurs only when identification and division are "ambiguously together."[33] Envisioning politics as a neat we/them or right/left overlooks this mediatory ground. In other words, if those on the opposing sides were completely opposed, there would be no common ground and no opportunity for rhetoric.

Further, neat right/left divisions gloss over the process by which individuals and societies develop political orientations. In rhetorical terms, we might understand political orientations in the Burkean sense of "orientation" or "piety," or as a process of putting together "what properly goes with what."[34] The themes or catchphrases used as prompts in neuroscience studies, or even the descriptors such as "liberal" and "conservative" themselves, can be considered idioi topoi, or special topics that have come to be associated with political

---

30. Mouffe, *On the Political*, 11.
31. Hawhee, *Bodily Arts*, 16.
32. Burke, *Rhetoric*, 25.
33. Hawhee, *Bodily Arts*, 16.
34. Burke, *Permanence and Change*, 74.

viewpoints over time. Some political topoi are shared across parties—including visual symbols (the American flag, the Statue of Liberty), songs ("God Bless America," "America the Beautiful"), colors (red, white, and blue), and other catchphrases such as "life, liberty, and the pursuit of happiness." Yet, each political party has also built up its own rhetorical topoi, each with its own history and fluctuations over time.

Further, while the Wilson-Patterson scale sought to free measures of political orientation from rhetorical appeal, we know from research on framing that even a single word can have a rhetorical effect; consider how the term *cold war* oriented political opinions during the 1980s.[35] Shortening sentences down to words cannot avoid rhetoricity. In fact, these shorts words and phrases simply function as topoi, or commonplaces of argument. We might think of topoi as tags or attributes that can be assembled, disassembled, and reassembled to make up what we call a political orientation. One's political orientation, then, can be understood as a shifting-yet-stable assemblage of topoi; yet topoi in turn can be understood as assemblages or, in Carolyn Miller's terms, as "a point in semantic space that is particularly rich in connectivity to other significant or highly connected points."[36] This makes it pretty much impossible to pinpoint what a particular topos means for a particular person at a particular place and time.

Take, for instance, legalized abortion. Using "abortion" as a marker of conservatism assumes, first, that the meaning of "abortion" is stable over time, and, second, that it means one particular thing (a particular law or practice). The Wilson-Patterson scale preceded Roe vs. Wade (which would legalize abortion in the US) by six years, but already in the 1960s public discourse about abortion had begun to shift. For decades, abortion was seldom discussed in public (although it was regularly practiced), but in the 1950s, as Celeste Condit recounts, this began to change as a result of many different factors. For one, doctors began to feel uncomfortable with their role in a procedure that was often legally ambiguous; they began a series of professional discussions on the topic.[37] These discussions led to a proposal for a "model law" developed by the American Law Institute in 1959; this law would permit abortions in cases of rape or incest, in cases of clear threat to the mother's physical or mental health, and in cases where a pregnancy "could be reasonably expected to produce a badly deformed or retarded child."[38] It is evident

---

35. Entman, "Framing: Toward Clarification of a Fractured Paradigm," 52.

36. Miller, "The Aristotelian Topos," 142.

37. Condit, *Decoding Abortion Rhetoric,* 23.

38. "New Proposals on Abortion," 4.

here that such framing used ableist language that portrayed disabled children as unworthy of life.

Meanwhile, in the popular media, lurid, graphic stories about illegal abortions began to circulate that featured women (either innocent young teenagers or married women) who were driven to abort a fetus due to extreme circumstances and provided shocking details about dangerous "back alley" abortions.[39] These stories, Condit argues, helped to shift public opinion in favor of legalization. Nonetheless, those who did not support abortion continued to push arguments that linked the terms *fetus* and *life,* often drawing on appeals to the authority of science and graphic visual images of aborted fetuses.[40]

In 1968, when Wilson and Patterson developed their scale, for some respondents "abortion" might have conjured up vivid images of these "back alley" abortions and sympathy for women damaged by them; for others, abortion may have drawn up images of mangled fetuses; still others might invoke religious concepts. Thus, while the term *abortion* seemed to Wilson and Patterson to have the advantage of reducing ambiguity inherent in sentence-length prompts, its meaning would be far from singular or clear.

Further, it is not clear that "abortion" would necessarily be associated with conservatism in a pure way. By 1967, 71 percent of Americans supported reform to the existing abortion law (which at the time allowed for abortions only in cases where the mother's life was in danger). This relatively high support suggests that at least some people who might otherwise be grouped as "conservatives" nonetheless were in favor of this change. Even today, while the number of Americans who consider themselves "pro-Life" (46 percent) or "pro-Choice" (47 percent as of 2016) is roughly equal, a much smaller percentage of people (just 19 percent in 2016) thought abortion should be illegal in all cases.

By examining polls further, we can see that support for abortion depends on the specific policy in question. Many more Americans support abortion when a mother's life is at stake (83 percent as of 2011), for instance, or when a pregnancy results from rape or incest (75 percent as of 2011) than when a mother is unable to financially support a baby (36 percent).[41] These key stipulations continue to form part of public debate about abortion, although the relative emphasis given to them shifts and new stipulations are added over time. With the availability of medical imaging, for instance, some have proposed that women should be required to see an ultrasound image of their fetus before proceeding with an abortion.

---

39. Condit, *Decoding,* 23–24.
40. Condit, *Decoding,* 61.
41. Gallup, "Abortion."

My point here is that "abortion" is not a stable signifier of conservatism or liberalism. It is not stable within a given time period, as it might connote different meanings for different people depending on their own exposure to arguments, stories, and information about that issue. It is not stable in its link to conservatism or liberalism, as certain components of abortion arguments may alternately gain or lose the middle ground or shift more to one side of the spectrum or the other. And "abortion" is certainly not stable across time or place but shifts as new arguments, stories, and policies are added to the general public repertoire of discourse.[42]

Despite these changes in meaning, abortion has at least stayed in public discourse since 1968 when Wilson and Patterson launched its scale, giving it at least some credibility as a marker of political orientation. Other topoi clearly stamp the original scale as a product of its time, such as "beatniks," "student pranks," and "jazz,"—all terms that would no longer register as markers of a political orientation.[43]

More broadly, what counts as "conservatism" or "liberalism" emerged in a specific historical period. For instance, "conservatism" itself emerged in the decades leading up to the development of the Wilson-Patterson scale. As Michael Lee explains in his book *Creating Conservatism*, a "canon" of political texts originating in the 1950s circulated key themes that now form the backbone of American conservatism, such as "freedom," "tradition," "fear of government," and "order." In Lee's terms, these kinds of themes constitute a political language, part of a "political community's standard argumentative stock, a communally specific collection of rhetorical resources in which ideological and symbolic diversity can yield potency, flexibility, and combustibility."[44] These terms become codes for managing community, identity, and history within a political group.

The topoi preferred by different political figures are readily identifiable from political speeches. Take, for instance, two different speeches from the 2016 election cycle: Hillary Clinton's Democratic National Committee speech and Donald Trump's Republican National Committee Speech. In the former, terms such as *family, economy,* and *health* dominate—all linked to the Democratic Party's emphasis on domestic policy issues.[45] In contrast, Trump's speech highlights *trade, immigration, violence,* and *opponents*—a set of topoi that together represent a collective sense (cultivated by Trump) of a nation at risk

---

42. For historical analyses of the rhetoric of abortion see Condit, *Decoding Abortion Rhetoric,* and Stormer, *Sign of Pathology.*

43. Wilson and Patterson, 266.

44. Lee, *Creating Conservatism,* 11.

45. "Hillary Clinton's DNC Speech."

and under attack by outsiders.[46] Certainly, these terms have not always been central to either party's political platform or with liberalism or conservatism per se; they instead reflect the sociohistorical context for contemporary politics. Thus, any order is, in Mouffe's terms, a "temporary and precarious articulation of contingent practices."[47] The sets of topoi that come to stand in for a political orientation represent not a "natural order" but a set of "sedimented practices" that can and do change over time.

Of course, the frameworks used in neuropolitics generally reflect the two dominant political parties in the US. Including different political groups would yield a broader array of terms and a broader set of choices. For instance, libertarians might add terms such as *rights, individual,* and *property;*[48] whereas Green Party candidates might constellate around terms such as *ecological, future, community,* and *local*—at least, according to the preamble and list of ten key values on their party website.[49] An analysis of the declaration of the Occupy Wall Street movement yields key terms such as *power, process,* and (in some overlap with the Libertarians) *rights.* In these documents, we also see terms not found in other platform statements, such as *colonialism* and *profit.*[50] In these sources, we see topoi that do not neatly overlap with those of the two major political parties in the US.

The diversity of political topoi identified thus far suggests that a simple right/left opposition is probably not tenable if we are to understand any kind of biological interactions with the sociorhetorical elements of political behavior. Further, the right/left opposition does not necessarily hold outside of democratic political systems. What sense would such a division mean in North Korea, Saudi Arabia, or Oman (to name a few countries where no political parties are permitted)? Take, for instance, key words that appear in a Saudi Royal speech—in the absence of a political platform or candidate speech to analyze. Here, key terms such as *security, cooperation, peace, kingdom,* and *unity* predominate.[51] The speech seems to contain within it a political model based on unification rather than division. In other words, in the absence of political parties that form identifications through agonism, in a monarchy it may be that the king or queen's rhetorical efforts focus largely on creating a single identification and on maintaining that identification by opposing the unified kingdom to an Other that lies outside the system. Sticking to a simple

46. "Full Text: Donald Trump 2016 RNC Draft Speech Transcript."
47. Mouffe, 18.
48. Libertarian Party, "2018 Platform."
49. Green Party, "Preamble"; Green Party, "Ten Key Values."
50. Occupy Wall Street, "Declaration."
51. Majlis Ash-Shura, "4th Term 3rd Year, 1428–1429 A. H."

right/left binary, then, does not accurately capture the full range of political affiliations that one might find across countries or even within a single nation.

To put these ideas together, we might say that political attitudes depend on a set of circulating themes, feelings, beliefs, icons, or, in rhetorical terms, topoi, that may coalesce around a particular label, such as "Christian conservative" or "The Resistance" (a term that arose in the wake of the 2016 election of Donald Trump) or hybrids such as "fiscally conservative, socially liberal." We might consider these labels as ways of summing up or tagging a cluster of topoi; in Burke's terms, they form a "general *body of identifications*."[52] These sets of terms may be organized in agonistic relationships (so that "conservative" is opposed to "liberal," but also that "fiscal conservative" and "social conservative" positions may also, at times, be in conflict). And we might see the same topoi being used within different clusters, albeit to signify different things or evoke different kinds of feelings. Topoi overlap and are pulled in different directions; rhetors try to push them this way or that.

Social and cognitive neuroscience studies grounded in a right/left metaphor binarize political orientations and overlook this fluidity. Choosing different metaphors might open up different lines of inquiry. The metaphor of the cloud captures the middle ground between stickiness and fluidity that characterizes political affiliations as clusters of topoi (and their accompanying practices, beliefs, and feelings). Clouds coalesce, but they also morph, drift, disperse, and reassemble over time. Clouds are temporal and spatial phenomena, which suggests another important aspect of political affiliation. How might one operationalize a cloudlike understanding of political orientations?

For one, researchers might develop metrics that reflect a wider array of political orientations and beliefs than ones based on a simple conservatism/liberalism distinction. They might draw on a more global understanding of politics to do so, recognizing the vast array of topoi and their relationships that appear across time and place. Even for research conducted within an American right/left paradigm, it would be interesting to question how participants respond to topoi that are not necessarily associated with either dominant perspective in this country, such as *unity, future,* or *local*. It might also be interesting to observe whether and how individuals may respond differently to those topoi depending on an imagined context or scenario. Just reading those words is much different from hearing them invoked in a political debate, speech, or news piece. More pressing in our current context might be studies of how individuals assess the veracity and trustworthiness of statements made by political candidates, how authoritarianism and demagoguery

52. Burke, *Rhetoric*, 26.

function, or how connections may be made across political differences. Social and cognitive neuroscience research might help to identify whether and how identification with an opposing viewpoint might occur, whether through narrative, emotional appeals, or appeals to common ground, all of which would be suggested by rhetorical scholarship.

Another line of research might be prompted by studies suggesting that our political affiliations reflect our own physical and temporal locations. For instance, membership in some churches strongly correlates with conservative party affiliation, particularly for congregations that value religious certainty, evangelism, fundamentalism, and a self-concept as socially marginalized.[53] While one might argue that individuals who are neurologically disposed to conservatism are simply more likely to join such churches, this explanation belies data suggesting that members of these churches are most likely to live in rural areas and represent lower relative socioeconomic status—unless one is willing to go so far as to argue that such people are in some way neurologically disposed to lower socioeconomic status, which would be a troubling claim indeed.[54] A more intriguing question might be whether certain kinds of habits (understood in terms of both locations and practices) coincide with certain ways of thinking. That is, do the patterns that neuroscientists find in studies of conservatism and liberalism evolve from habits of thought? We might find evidence of those habits in what people hear in churches, what they read, the media they consume, the local stories that circulate at coffee shops and church socials, and even the unarticulated social norms of dress and behavior in such communities. For instance, how might individuals respond to a given prompt before and after hearing a sermon (even a fairly neutral one)?

We might ask similar questions about other locations. For instance, Republican Party affiliation correlates with suburban sprawl—specifically with newer housing stock and increased reliance on cars.[55] While self-selection may account for this pattern to some extent, social policy researcher Thad Williamson finds that even those who would like to live elsewhere tend to have more conservative attitudes than their urban peers. Williamson suggests that living in a suburban community may lead individuals to privilege lower taxes because they already pay for many services through homeowner fees (such as security, parks, and recreation facilities), whereas individuals in urban communities might be more likely to see the value of publicly funded services.[56] Because they live in relatively enclosed (if not gated) communities, suburban

53. Wald, Owen, and Hill, "Political Cohesion in Churches," 207.
54. Wald, Owen, and Hill, 207.
55. Williamson, "Sprawl, Spatial Location, and Politics," 915–16.
56. Williamson, 913.

individuals may be less likely to experience diversity directly. Their experience of "public" differs, as Susan Bickford has argued, from those of city dwellers, who regularly encounter strangers as soon as they step out their front doors. These daily experiences—the actual presence of other people—Bickford argues, shape political beliefs "in terms of the awareness of different perspectives that must be taken into account in forming opinions."[57] What might happen, then, if prompts in a scientific study come after immersing individuals visually and aurally in a "public" scene (a town square, for instance) as opposed to a "private" scene (a suburban cul-de-sac)?

Perhaps most troublingly, social and cognitive neuroscience studies of politics typically exclude considerations of race and gender. In this way, they universalize political differences and overlook how political party affiliation tracks with race, gender, and ethnicity. Yet polling data and other research studies show that race and gender significantly affect voting patterns and political affiliations. For instance, women—especially minority women—disproportionately vote for Democratic candidates.[58] (Support for conservative candidates by white women remains high, as of 2017, although not as high as for white men.) This has not always been the case. In the 1960s women were actually more supportive of conservative views. In the 1970s, though, this pattern shifted, as men and women were equally likely to vote for Republicans or Democrats in the 1972 and 1976 presidential elections.[59] By the 1980s women became more likely to support Democrats. This dramatic shift cannot be accounted for by genetics or biology. Clearly, though, dramatic social changes occurred during this period, namely the feminist movement of the 1960s and 1970s. Whereas earlier "neither party actively championed women's rights,"[60] in that period both political parties jockeyed for women's support by including women's issues in their platforms.[61] (It was during this period that women were equally likely to support Democrats or Republicans.) However, by 1980 the Republican Party had jettisoned support for women's issues, especially the Equal Rights Amendment. The Democratic Party became the only party actively championing women's rights, and support from women followed suit.[62] This evidence suggests that support for political parties depends heavily on how each party actively courts different groups of voters.

---

57. Bickford, "Constructing Inequality," 370.
58. Chaturvedi, "A Closer Look at the Gender Gap in Presidential Voting."
59. Chaturvedi.
60. Wolbrecht, *The Politics of Women's Rights*, 5.
61. Wolbrecht, 5.
62. Wolbrecht, 6. See Wolbrecht for a deeper analysis of how these positions evolved.

In the US, racial identifications also predict political orientations. Overall, white Americans lean right—particularly white evangelical protestants and white southerners.[63] In contrast, people from other racial and ethnic groups (including African Americans, Asian Americans, and Latino/a Americans) lean left. If political orientations were largely influenced by genetics, we would not expect to see such obvious racial divides—unless we were willing to support an idea of innate racial differences in neurobiology. Given the problematic history of linking minority groups to brain differences, I hope that few would be willing to support that idea. The differences in racial support for political parties are more likely due to the ways that those parties appeal to different groups of voters and their interests and have done so historically, over time.

A key line of research in neuropolitics could involve a more diverse group of study participants. Even when researchers are not explicitly studying gender and racial differences in political orientations, study participants should reflect the diversity of the electorate. Given the tendency for researchers to draw their pools from university communities, this may not often be the case. (Often, researchers don't provide a breakdown by race.) Ensuring diverse representations among study participants and accounting for differences, if any, seems crucial, but results of studies regardless of participant populations should at least acknowledge how political orientations are distributed at the population level across racial and gender groups. The idea is not that "black Republicans" might respond differently to study prompts from "white Republicans" or "black Democrats" (that approach assumes neuroessentialism) but that the unequal distribution of political orientations across demographic groups suggests that something more than biology is playing a role.

The question of race raises another challenge with how neuroscientific studies currently operationalize politics. Key terms used in metrics such as the Wilson-Patterson scale may unwittingly index racist beliefs, even when they seem neutral. More broadly, political orientations themselves track with views about race. Thus, studies that purport to measure how participants respond to certain kinds of prompts—such as words like *immigration* or *patriotism*—may actually be measuring racism as much as they measure political beliefs.

Indeed, American politics is awash in terms that subtly appeal to white Americans' fears of the Other. Rhetoric scholar Jeremy Engels argues that the political climate in America is driven by a "politics of resentment" that "works to capture popular resentment and direct it away from structures of oppres-

63. Pew Research Center, "A Deep Dive Into Party Affiliation."

sion and toward our neighbors and fellow citizens."[64] While resentment may be endemic to democratic societies, Engels notes, it was historically deployed to unify the American people by directing negative feelings to external enemies. In the middle of the twentieth century, however, political discourse began to instead create divisions within the American polis, and these divisions relied specifically on coded appeals to race. As Engels explains, the Republican Party in this era began to rely on white fear of blacks and their resentment of liberal social policies in order to appeal to the white voters newly constituted as the "silent majority" or "Middle America."[65]

This approach grounded Richard Nixon's campaign, which came to be known as the Southern Strategy. The scheme was outlined in Kevin Phillips's 1969 book, *The Emerging Republican Majority,* where Phillips openly calls Nixon's success in 1968 a "repudiation" of the Democratic Party for "its ambitious social programming, and inability to handle the urban and Negro revolutions." Phillips shrewdly noted that "ethnic and cultural animosities and divisions exceed all other factors in explaining party choice and identification."[66] As Nixon's advisor, he developed a strategy that appealed to white resentment without explicitly invoking race.

In the ensuing years, conservatives continued this strategy of "colorblind rhetoric," or a set of discourses that seek to restrict progress for people of color but are masked by terms such as *"fairness, open access, and equal opportunity."*[67] For instance, Ronald Reagan deployed similarly coded language when he spoke repeatedly about welfare throughout his career. In 1986, for instance, he gave a radio address to the nation on the topic of welfare reform. The speech focused on a "family crisis" that Reagan located in "the welfare poor, both black and white." Despite this caveat, however, the speech is peppered with terms that can easily be understood as playing upon racial stereotypes. The crisis he refers to is "concealed behind tenement walls or lost in the forgotten streets of our inner cities"; he refers to teenaged mothers whose fathers "are often nowhere to be found." Indeed, he states, "in some instances you have to go back three generations before you can find an intact family." The intact family, in contrast, provides "the courage, willpower, and sense of security that have enabled millions of Americans to escape poverty and grab hold of the rungs on the ladder of opportunity." Here he begins to fill in other white-coded terms—"strength of character," "hard-working," "success." Reagan goes on to pin the "family crisis" he describes on welfare pro-

---

64. Engels, *The Politics of Resentment,* loc. 1282–83, Kindle.

65. Engels, loc. 1430, Kindle.

66. Phillips, *The Emerging Republican Majority,* 18.

67. Holmes, "Affirmative Reaction," 26.

grams that he claims incentivize young mothers to avoid marriage in order to claim higher benefits.[68] These terms invoke racialized political identities, such as that of the "welfare queen," which as Ange-Marie Hancock explains, are designed to invoke "stereotypes and moral judgments" of others (typically on the basis of race, class, and gender) against which individuals can constitute their own political identities.[69] When researchers employ topoi such as *fairness* or *opportunity*, then, they may be unwittingly using racially coded terms to evoke responses in their participants, playing into internalized racist stereotypes rather than genetic predispositions to a certain political orientation.

Most recently, we see Trump harnessing white resentment with key terms like *immigration* and *make America great again*. When he announced his candidacy on June 16, 2015, Trump immediately set the stage for his campaign by playing on white Americans' fears of foreigners (the Chinese and Japanese, in particular) and immigrants (particularly Arabs and Mexicans, the latter which he notoriously described as bringing drugs, crime, and rape to America).[70] Trump presented a vision of America as threatened externally ("Our enemies are getting stronger and stronger by the way, and we are getting weaker") and internally, doubling down on the racialized politics of resentment that drove the political discourse of Nixon and Reagan. Thus, a topos such as "immigration" cannot be implemented in a neuroscientific study without recognizing its implicit connection to racism.

Even issues that may not seem to have anything to do with race in fact track with racial attitudes, including gun control, the death penalty, and health care reform.[71] To take gun control as one example, whites are twice as likely as blacks to own guns for reasons that include "personal safety" and "self-protection." These fears for personal safety correlate with fears of black violence and negative attitudes toward blacks in general.[72] Indeed, O'Brien et al. suggest that white Americans who demonstrated the most bias against African Americans were also more likely to oppose gun control measures. In a rhetorical sense, then, a term like *gun control* masks a host of attitudes and beliefs that are racially charged. In the US a host of problems are coded as "race" problems, according to political science scholar Danielle Allen, including problems related to "welfare, employment, crime, drugs, [and] gangs," and so are issues that "require implementation across race lines" such as "health care, abortion,

---

68. Reagan, "Radio Address to the Nation on Welfare Reform."
69. Hancock, *The Politics of Disgust*, 15.
70. Trump, "Our Country Needs a Truly Great Leader."
71. O'Brien et al., "Racism, Gun Ownership and Gun Control," 23.
72. O'Brien et al., 1–2.

housing and real estate, city planning, [and] public education."[73] By failing to address the tremendous impact of race on American political discourse and orientations, neuroscience studies participate in a "color-blind" rhetoric that masks the workings of racialized resentment in shaping political orientations and the extent to which those resentments have been deliberately manipulated to carve out political identities.

More broadly, the fear responses that seem to be present in right-leaning individuals are not simply biological markers but potentially also markers of a racialized social history. As Allen puts it, "Social or economic loss becomes political when citizens believe themselves disadvantaged by a collective decision. Regardless of whether their beliefs are reasonable, they will be registered in negative emotions like anger, resentment, disappointment, and despair."[74] We see this politics of fear and distrust at play currently in current rhetorics on the Right (not just the "Alt-Right"), which position "Middle America" (and often white men, in particular) as disadvantaged by current social and economic policies (from welfare to health care). The racialization of American politics, then, suggests factors at play that are not easily accounted for by social and cognitive neuroscience studies that assume a simple right/left distinction without factoring in race.

By working with humanities researchers who study politics and raveling together insights from those fields, researchers in the social and cognitive neurosciences might more generally contextualize their results in introduction and discussion sections of articles. Here, they might draw on research that recognizes the limitations of neuroscience studies that overemphasize fixity in political orientations and underemphasize their variability and interconnectedness with other social and contextual factors.

More pressingly, by focusing on neurological traces of political orientations, researchers may contribute to a relatively depressing view of future political discourse. This view is espoused by political scientists Marc Hetherington and Jonathan Weiler in their book *Authoritarianism*, which draws on neuroscience research to argue that "the colliding conceptions of right and wrong embedded in the opposite ends of this continuum make it difficult for one side of the political debate to understand (perhaps, in the extreme, even respect) how the other side thinks and feels."[75] They ground this problem in biology, suggesting that "since automatic processes in the brain likely govern these opinions and behaviors, it might not be possible for people to

---

73. Allen, *Talking to Strangers*, xv.
74. Allen, 46.
75. Hetherington and Weiler, *Authoritarianism and Polarization in American Politics*, 5.

understand the moral code that their adversaries are living by."[76] In short, a view grounded in neurobiology like this one tends to yield a pretty nihilistic outcome, in which liberals and conservatives will never find common ground, political debate will become increasing polarized, and, ultimately, authoritarian regimes will rise to power.

Despite these limitations, researchers do suggest that cognitive political orientations may be subject to change. Hetherington and Weiler suggest that during times of great political upheaval or threat—as in the years following 9/11—those who are less authoritarian can be activated "toward the use of instinct over cognition, which the more authoritarian apparently rely on commonly."[77] While troubling for anyone who worries about the implications of authoritarianism, this perspective suggests that political orientations may not simply be stable cognitive styles but may be situationally specific. Hetherington and Weiler do not delve deeply into this question. It may be the case, though, that this cognitive style can be both: perhaps those who are conditioned to think in black-and-white terms, to react to threats to law and order—in part through repeated exposure to such rhetoric—strengthen those neural networks. For others, that particular network may remain latent but can either be activated temporarily or, over time, strengthened under continued exposure to rhetorics of fear and threat. And if political orientations are not innate, but *conditioned,* there is a possibility that they may be conditioned differently. If instinct over cognition can be situationally enacted, can the opposite as well?

In the neuroscience studies I examined, there were a few hints that this could be the case. Schreiber et al. note that the direction of causality in their study is not entirely clear. Does a "conservative" brain structure lead one to become conservative, or does one's conservative acts and behavior shift brain structures? In other words, the authors ask whether "acting as a partisan in a partisan environment may alter the brain."[78] In their study, Kanai et al. note that current studies do not clearly indicate whether "the changes in brain structure that we observed lead to changes in political behavior or whether political attitudes and behavior instead result in changes of brain structure."[79] These lines are provocative because they suggest that the polarization of political debate is not the only possible course of action and not simply the outcome of some natural biological process.

---

76. Hetherington and Weiler, 198.
77. Hetherington and Weiler, 110.
78. Schreiber et al., 3.
79. Kanai et al., "Political Orientations Are Correlated with Brain Structure in Young Adults," 679.

Presumably, a different kind of political rhetoric could potentially yield different kinds of cognitive capacities, such as an ability to weigh both sides of an issue, to avoid confirmation bias, and to work collaboratively with others to develop solutions to a problem. These capacities are well recognized by ancient rhetoric scholars who stressed rhetorical training as a "gymnastic for the mind,"[80] clearly invoking the effect of rhetorical training on the mind (or what we would now likely term the brain) that is akin to the effect of gymnastics training on the body. Robert Terrill describes the goal of rhetorical education as "something more than the training of tongues; we are engaged in the formation of citizens."[81] Engels argues that "without an education in rhetoric, democracy will remain nothing more than an unrealized possibility haunting our dreams."[82] In particular, the Mt. Oread Manifesto on Rhetorical Education outlines some of the particular skills that a rhetorical education should provide, including the ability of "average citizens" to "perform rhetorical analyses of the discourse around them," to "advance an idea in the public sphere and engage in meaningful deliberation about ideas," and to "recognize the limits and possibilities of a given mode of communication."[83] To develop what Quintilian termed *facilitas,* or "the capacity to produce appropriate and effective language in any situation,"[84] would certainly require an intensive course of study, as the Mt. Oread Manifesto notes: "the *paideia* is a life-long pursuit that habituates citizens to rhetorical practice from kindergarten through adult education."[85] This language of habituation evokes an understanding of the brain and mind as flexible, and subject to change over time—a view that is not inconsistent with neuroscientific concepts of brain plasticity but one that is not generally well represented in studies of neuropolitics.

This view of rhetorical skill stresses rational thought, but others focus on other capacities. In particular, we might argue that rhetoric offers a way to cultivate trust by "inject[ing] friendship into citizenship." Allen locates this capacity in Aristotle's *Rhetoric,* which she usefully describes for the uninitiated as "neither a guide to manipulation nor a superficial manual of style, but rather a philosophically subtle analysis of how to generate trust in ways that preserve an audience's autonomy and accord with the norms of friendship."[86] To do so, however, we might need to move beyond the Aristotelian framework

---

80. Isocrates, "Antidosis."
81. Terrill, "Mimesis, Duality, and Rhetorical Education," 296.
82. Engels, loc. 233–34, Kindle.
83. "The Mt. Oread Manifesto on Rhetorical Education 2013," 3–4.
84. Murphy, *A Short History of Writing Instruction,* 19.
85. "The Mt. Oread Manifesto on Rhetorical Education 2013," 3.
86. Allen, 141.

that Allen outlines. This framework focuses primarily on acts of persuasion but does not address the key skill that Krista Ratcliffe defines as "rhetorical listening," or "a stance of openness that a person may choose to assume in relation to any person, text, or culture," a stance that is particularly useful in cross-cultural exchanges.[87] While neuroscientists might argue that some individuals simply possess this stance as a cognitive style—an openness to considering what others have to say—a rhetorical scholar might argue that this stance can be practiced in classrooms and modeled in public discourse. Together, though, the rhetorician and the neuroscientist might investigate whether such practice might lead to neurological changes. Over time, does someone who practices rhetorical listening develop greater tolerance for ambiguity? Through practice, can one diminish or override the automatic fear response that may drive authoritarian thinking?

As Crowley argues, ancient rhetoricians recognized the powerful role of emotions in changing beliefs;[88] certainly, Aristotle's *Rhetoric* devoted considerable time to outlining appeals to emotions, and Quintilian described the effectiveness of *enargeia,* or vivid description, for its ability to evoke emotional responses. Contemporary neuroscience research might enrich traditional rhetorical education by prompting us to consider more deeply the embodied and affective nature of public deliberation and politics. In their review of the literature, Kanai et al. note that conservative political orientations are associated with a larger amygdala and greater sensitivity to fear, and with greater sensitivity to disgust (typically associated with the insula). In support of this idea, some recent neuroscience research concurs that political beliefs, in Dodd et al.'s words, may become "biologically instantiated in a fashion that renders them sticky and slow to change—somewhere between wholly static and completely malleable."[89] And Smith et al. acknowledge that it is not "that biology causes politics or that politics causes biology but that certain political orientations at some unspecified point become housed in our biology, with meaningful political consequences."[90] If interpreted carefully, these and other studies lend credence to the growing (or perhaps renewed) recognition that rhetoric is embodied, ambient, material, affective (a topic I will explore in more detail in the following chapter).

In her review essay on critical affect studies, Jenny Rice describes how affect offers us a way of understanding rhetoric (especially political rheto-

---

87. Ratcliffe, *Rhetorical Listening,* 1.

88. Crowley, *Toward a Civil Discourse,* 79.

89. Dodd et al., 647.

90. Smith et al., "Disgust Sensitivity and the Neurophysiology of Left-Right Political Orientations," 8.

ric), as additive and associational, involving articulations or networks of discourses, affects, and emotions.[91] Rice gives the example of anti-immigration discourse on websites, where posters to an online forum might link foreign-language signs in their neighborhood with claims about increased crime. These two ideas are not logically connected, but they both belong to a web of images, affects, and ideas that circulate around the topic of anti-immigration: the idea that immigrants bring crime. Instead, Rice argues, individuals are already swimming in a pool of interconnected ideas and images about immigration, and they might engage within that discourse by adding ideas or making new links among them. Rhetorical education, then, should help students not simply to invent arguments whole cloth but to identify ways of disconnecting and reconnecting ideas, making new links that can shift what Crowley calls ideologics by appealing not only to reason but also to affects, emotions, and visceral responses.

Any educational program will "change" the brains of those who engage in it, so it is meaningless to say that rhetorical education might do so. However, it might be interesting to consider whether and how rhetorical education can produce specific cognitive changes that align with the democratic goals with which we often link it. Clearly, there are interesting ways to think through the intersections between rhetoric and neuroscience. I take up some of these connections more fully in the following chapter, where I consider the extent to which neuroscience studies of affect can inform rhetorical study, and vice versa.

---

91. Rice, "The New 'New,'" 210.

CHAPTER 6

Neuroaffect

IN 2016 the Tragically Hip launched a coast-to-coast tour of Canada. It would be the last for the band, which had been a fixture on Anglo-Canada's radio stations, as the opening montage for Hockey Night in Canada,[1] in bars, and around cottage campfires for over twenty-five years. The Hip's lead singer, Gord Downie, had been diagnosed with terminal brain cancer. It was what we might call an affective event. Canada's Prime Minister, Justin Trudeau, appeared at the final concert of the tour, in Kingston, the Hip's hometown, wearing a Hip T-shirt with his "Canadian tuxedo" (blue jeans and a matching blue denim jacket). Fans posed for selfies with Trudeau in the street as they waited to go in to the concert. The concert itself was televised in its entirety, ad free, on the CBC. People organized viewing parties. Canadian fans of the Hip gathered in bars across the nation to watch the final concert, alternately singing, dancing, crying, and cheering throughout the nearly three-hour concert. Others gathered online, sharing their reactions in real time through social media. The mood was alternately celebratory, sad, triumphant, bittersweet.

How can rhetoric and neuroscience make sense of this type of event? In this chapter I draw on theories of affect to suggest that "classical" models of emotion and affect do not fully account for this type of event. First, I show

---

1. Battle, "Remembering Tragically Hip's Contributions to Hockey Night in Canada."

how rhetorical theories of affect that draw on neuroscience and psychology often split off affect from discourse. Then, I show how this pattern reflects a classical model of affect and emotion that has historically informed psychology and neuroscience. Next, I suggest that more recent neuroscience and psychology research on affective practice and constructed theories of emotion offer a more productive model, one that also accords better with rhetorical perspectives. This view, which considers affect, emotion, language, and culture to be raveled together, provides one example of how humanistic theories can productively merge with neuroscience to develop richer understandings of human cognition and behavior.

HOW RHETORIC STUDIES AFFECT

It would be tempting to interpret the final concert of the Tragically Hip as a moment of unmediated, affective response to an arousing multimodal stimulus. In the humanities, those who draw on the affect theory of Brain Massumi, in particular, often view affect as separable from or prior to discourse. In his article "The Autonomy of Affect," Massumi draws on a set of experiments to posit intensity as a "nonconscious, never-to-conscious autonomic remainder" that is "outside expectation and adaptation, as disconnected from meaningful sequencing, from narration, as it is from vital functions."[2] This view of intensity not only differentiates affect from emotion and discourse but also differentiates particular types of autonomic responses. In particular, based largely on a study by Hertha Strum from the 1980s, Massumi privileges the skin as the site of autonomic response that is "faster than the word."[3]

In this study, Strum asked a group of children to view a short film in three different test conditions. In one, the film was shown without any narration. In the second condition, the film was shown with narration that factually related the events in the film. In the third condition, the film was shown with narration that also added emotional content. The researchers monitored participants' physiological reactions using three measures: heart rate, respiration, and skin resistance. They found that the children reacted most strongly to the second test condition, with factual narration, but that they found this condition less pleasing than the other test conditions. Notably, in the study, the factual condition produced higher heart rate, higher respiration, and lower skin resistance. Massumi takes this to mean that skin resistance

---

2. Massumi, "The Autonomy of Affect," 86.
3. Massumi, 86.

was somehow operating on a different track from the other two measures. He argues that heart rate and respiration link to the content and meaning of the film but suggests that skin conductance was somehow not connected to content. He accordingly labels this last condition as being equated with nonsymbolic intensity and therefore affect. However, lower skin resistance actually means *higher skin conductance*—and therefore higher intensity. That is, the three measures were not in fact working at odds with one another. As Ruth Leys has explained, Massumi (through either willful misreading or ignorance) misinterprets the term *skin resistance* and then uses it to draw this conclusion about the unqualified, noncognitive nature of affect.[4] For our purposes, though, what is important is that, rather than operating completely independently from the content of the film or from the other two measures of arousal, the skin was responding alongside the other physiological measures and alongside the narrative content of the film. Nonetheless, Massumi uses this study to ground a view of affect that renders it separable from, and prior to, emotion, language, and cognition.[5]

It is beyond the scope of this chapter to work through the entire corpus of humanistic inquiry into affect, and I do not mean to paint all theorizations of affect with broad strokes. Here, though, I hope to show that at least some strands of humanistic work on affect share a similar set of assumptions to Massumi: that affect is separable from language, thought, and emotion; that affect is a response to certain stimuli; and that affect is therefore universal but can (despite its lack of referentiality) somehow also ground cultural and rhetorical analysis of events and experiences.

To begin, take Melissa Ames's article "Exploding Rhetorics of 9/11: An Approach for Studying the Role That Affect and Emotion Play in Constructing Historical Events." Here, Ames posits a model called the "affect continuum" in

---

4. Leys, "The Turn to Affect: A Critique," 448.

5. Recent views of emotion that seek to explain the relationship between language, affect, and emotion might also shed light on this study, as I show in the next section. Strum found that, when listening to narration that included emotional information, participants had lower autonomic responses. This finding could be in line with studies that suggest that labeling emotions actually dampens autonomic responses. This is the same principle behind the advice given to parents to encourage children to "use their words" when they are upset—and it is the same reason, perhaps, why young toddlers who lack the words to accompany emotions have the most intense tantrums. Thus, while Massumi (and some who draw on his work) wants to keep affect separate from emotion and from cognition, these more recent studies seem to suggest that the three are raveled together. (This is not to say that they are always *tightly* raveled, that there is no such thing as an autonomic response that is beyond signification or consciousness; instead, I'm suggesting a more dynamic, interactionist model wherein, in a given situation, affect, emotion, and cognition *may* be working together. This idea, I think, squares with emerging neuroscience research but is of course subject to refinement and change over time.)

which a stimulus is introduced and followed by a first-order, instantaneous, intense bodily response, followed by a second-order emotional response, and then by a third-order, more lasting "structure of feeling" or "mood."[6] This model grounds the analysis of affect in the article. Here, the idea is that bodily affects lead toward cultural "structures of feeling" in a unidirectional manner, at least according to the model she lays out. She concludes that there exists "a sort of shared, collective, national state of affect—a cultural 'mood' that seemed to linger more than a decade after the events of that day."[7]

In contrast, let me share a personal anecdote. On the morning of September 11, 2001, I walked from the house I shared with several other second-year master's students straight to my first-year composition class. Right as class started, a student raised his hand and said that a plane had hit the World Trade Center. I think my class must have started at 9:00 a.m., just minutes after the attacks. Yet I was *unaffected* by his words (or, as Jenny Rice might put it, my affective orientation was one of indifference or apathy). In my mind, I pictured a small plane, maybe a Cessna, and a single attacker who probably died in the crash. I think I said, "Oh, really?" and went on with the class. Afterwards, I walked through the student union on campus and saw the now-iconic video of the attack on the TV screens. Groups of students were huddled around. The students seemed to be participating in a collective, affective moment of outright terror, fear, and sadness mingled together. Yet I felt that I was somehow missing something. Intellectually, I knew this was big deal, and I knew that thousands of people were feeling scared, anxious, terrified. But as a Canadian, I didn't understand the symbolism of the World Trade Center. I didn't feel as though I, *personally,* or as a citizen of a nation, was being attacked. At the time, I had been living in the US for barely a year. I did not belong to the same affective *or* symbolic community as those around me.

If we take Ames's model, my experience does not make sense. In her version, the stimulus (in this case, images of the World Trade Center attacks) would produce, first, an automatic visceral response, followed by an emotional response, and then by a collective "structure of feeling." But without the cognitive or symbolic knowledge that the American students shared, I didn't feel the same visceral response (at least, I don't think that I did—since we don't go around with skin galvanometers attached to our bodies at all times, it is hard to say). But I think we have at least some sensitivity to these kinds of responses—the feeling that you've been "punched in the gut" when you receive bad news, or the goosebumps that appear when hearing a particu-

---

6. Ames, "Exploding Rhetorics of 9/11," 181.

7. Ames, 185.

larly compelling musical performance. I would argue these affective responses are culturally inflected, at least to some degree. They depend on symbolic knowledge but also on what Kenneth Burke would call systems of piety, the *integration* of the bodily, affective, emotional, and symbolic that make up our habituated ways of living. In this view, affect, emotion, language, and culture feedback on each other and also feed forward.

We see a similar tendency toward universalizing affective response in Brian Ott's "The Visceral Politics of *V for Vendetta*." In this article, Ott argues that the film in question "enlists and mobilizes viewers at a visceral level to reject political apathy and to enact a democratic politics of resistance and revolt against any state that would seek to silence dissent."[8] Like Ames, Ott relies on theories of affect that separate it out from interpretation, discourse, or emotion. He writes, "Since the figural can only be felt or experienced, rather than read or interpreted (like discourse), the rhetoricity of cinematic figures is best approached on an affective register."[9] Affect, he argues, refers to direct sensory experiences and to "the feelings, moods, emotions, and/or passions they elicit."[10] Thus, he is using the same type of model in which affect is nonsymbolic and arhetorical but nonetheless leads to feelings or emotions. While Ott provides one caveat, recognizing that not all viewers will respond to the film in the way he outlines due to "personal politics, background, and previous rhetorical experiences," he maintains that the film nonetheless *invites* a certain kind of response. The task of the rhetorical critic, then, is to "show how a particular rhetorical experience works" regardless of whether it affects everyone in the intended manner.[11] Here again, we have a model that posits bodily affects to be the primary mode of response, followed by emotions or feelings and only later by linguistic or cognitive operations. Because the affective register is posited as embodied but not linguistically shaped, it can therefore be put forward as having some kind of universal appeal that the rhetorical scholar can identify and uncover. Admittedly, Ott does provide some openings for feedback from previous experiences, noting that the bodily responses to a film include recall of "previous cinematic and non-cinematic experiences, which in combination evoke affective responses,"[12] but he does not explain how this kind of feedback affects his basic model of affect which, following affect theorists, is nonetheless somehow prior to and separate from interpretation.

8. Ott, "The Visceral Politics of *V for Vendetta*," 40.
9. Ott, 42.
10. Ott, 42.
11. Ott, 48.
12. Ott, 49.

Similar discrepancies emerge in Josue David Cisneros's "Looking 'Illegal': Affect, Rhetoric, and Performativity in Arizona's Senate Bill 1070." In this article, Cisneros first argues that "affect does not take shape entirely separate from representation," since "discourses participate with embodied experience, public culture, and historical memory to articulate affinities and emotional investments."[13] He uses this understanding of affect to demonstrate how Latino/a immigrant bodies are posited as *affectively marked* in anti-immigrant legislation such as SB 1070, such that illegal status can purportedly be recognized by embodied features (such as appearance and accent) as well as by embodied responses of fear on the part of non-immigrants. While his model seems to assume the entanglement of bodies, affects, cultural meanings, and symbolic representations, though, Cisneros nonetheless remarks in a footnote that (per Massumi) he understands "affect as a mode of perception that is prior to cognition and that is dispersed throughout the body in a visceral state."[14]

Thus, even when rhetorical critics seem to want to move toward a more dynamic, interactive model of affect and its relationship to rhetoric, emotion, and culture, we still sometimes find a residual (and sometimes overt) reliance on a model that assumes a straightforward, linear stimulus/response pattern; that separates affect out from emotion, language, cognition, and culture (thereby duplicating a mind/body division); and that can lead to overgeneralization and universalization of responses. As I show in the next section, this model draws from one that has been dominant within psychology and neurosciences for decades but one that, of late, has been called into question by researchers positing a more interactionist model in which language plays a key role.

HOW NEUROSCIENCE STUDIES AFFECT

Historically, neuroscience studies of affect and emotion shared many of the same qualities as rhetorical studies. Here I will ravel back, identifying the metaphors that have informed traditional studies of neuroscientific affect and emotion. These metaphors have constituted affect/emotion to be innate, precultural and prelinguistic responses that could therefore be studied using animal models using stimulus-response tests. Each emotion, it was hypothesized, would correlate to a specific brain region or network. Ironically, how-

13. Cisneros, "Looking 'Illegal,'" 137.
14. Cisneros, 150 fn2.

ever, while humanists have drawn on "affect" in order to usher in a turn away from the so-called linguistic turn that relied heavily on social constructionism, neuroscientists have recently turned *toward* linguistic and constructionist understandings of emotion and affect. In these more recent conceptualizations, emotion/affect are intertwined with language and culture.

For many years, neuroscientists drew on an understanding of affect and emotion that situated both outside of language and culture, instead understanding them as universal, built-in responses to environmental stimuli. We can see this understanding of affect and emotion at work in the 1956 article that James Olds, a psychologist from McGill University, published in *Scientific American,* called "Pleasure Centers in the Brain." The first paragraph of the article, alone, demonstrates some of the ways that earlier brain metaphors percolate up through history:

> The brain has been *mapped* in various ways by modern physiologists. They have located the sensory and motor systems and the *seats* of many kinds of behavior—*centers* where *messages* of sight, sound, touch and action are received and interpreted. Where, then, dwell the "higher feelings," such as love, fear, pain and pleasure? Up to three years ago the notion that the emotions had specific *seats* in the brain might have been dismissed as naïve— akin perhaps to medieval anatomy or phrenology. But recent research has brought a surprising turn of affairs. The brain does seem to have definite *loci* of pleasure and pain, and we shall review here the experiments which have led to this conclusion.[15]

Here, we see earlier metaphors of seats and centers (see chapter 1) combined with those of communications (messages) and mapping. Notably, Olds here makes the connection to the earlier study of phrenology explicit, even as he acknowledges its seeming naiveté.

By implanting electrodes into rats' brains, Olds could directly stimulate brain regions and note how it affected the rats' behavior. Then, Olds could postulate how specific brain regions correlated with specific behaviors. His most famous experiment, conducted with Peter Milner in 1954, involved placing rats in what was called a Skinner box or "operant conditioning chamber," created by B. F. Skinner. The box included, in its most basic form, a lever that a rat could press to obtain a reward. Traditionally, the lever would produce a food pellet or similar item. The lever would also be connected to a switch so that it would be possible to record the number of times it was pressed. Olds

---

15. Olds, "Pleasure Centers in the Brain," 105; emphasis added.

and Milner rigged their Skinner box so that the lever would directly stimulate the rat's brain. After a training period, they noted how often rats would press the lever when different areas of the brain were stimulated. In their 1954 experiment, they found that some rats would press the lever up to 742 times an hour. Those rats were found to have electrodes in "reinforcing structures," such as the nucleus accumbens. Olds and Milner argued that, based on this study, they had "perhaps located a system within the brain whose peculiar function is to produce a rewarding effect on behavior."[16]

Olds's experimental setup *literalizes* the metaphors of location and circuit. By implanting an electrode in a specific location of the rat's brain, and then connecting the rat's brain's circuitry to an externalized circuit via an electrode, Olds claimed to be simply measuring a reaction that was already there. From another perspective, however, what is actually being measured here is not simply the rat's brain but the rat's brain as part of a circuit or system that includes the cage, treadle, electrodes, experimenters, and so on. The line between the metaphor of the seat/circuit and the actual seat/circuit is obliterated, as is the line between the circuit in the rat's brain and the circuit created by the electrode planted within it and the treadle that stimulates it; the metaphor produces reality rather than merely reflecting it.

Olds's rats behaved much like a drug addict. Rather than fulfilling a drive or need, they seemed to self-stimulate entirely for pleasure—up to 7,000 times an hour, in a 1958 experiment.[17] Olds noted:

> If the electrode was placed in the brain at a point at which maximum self-stimulation is produced, the rat, after its very first electric stimulation, began to search and pursue eagerly . . . After the rat had pressed the pedal a second or third time, it ceased to wander and began to respond at the rate of one or two pedal-presses per second.[18]

Part of Olds's task was to overturn the view that emotions emerge from a lack. His rats were not in a deficient state when they stimulated the brain lever; instead, pleasure was for them a positive flow of affect. His rats demonstrated that view nicely, since they couldn't possibly "need" to stimulate their brains 7,000 times an hour. Once he had established that pleasure exists in the brain, Olds used similar methods in a series of follow-up articles, refining his experiments to pinpoint brain regions more accurately and to clarify what psycholo-

---

16. Olds and Milner, "Positive Reinforcement Produced by Electrical Stimulation of Septal Area and Other Regions of Rat Brain," 426.

17. Olds, "Self-Stimulation of the Brain," 316.

18. Olds, "Self-Stimulation of the Brain," 316.

gists meant by "reward." Olds argued that his experiments helped to disprove the "classical theory" of reward as the reduction of pain or fulfillment of a drive; instead, he argued, "there are anatomically separate mechanisms for reward and punishment in the brain," which means that the two phenomena must be separate as well.[19] As a rhetorical device, the idea of a reward center helped Olds make important arguments about the *nature* of reward, not just its localization in the brain, and the nature of reward in general, not just the nature of reward in rats. The ability of an experimental operationalization of emotions on animals to stand in for emotion in humans (or emotion writ large) represents another form of neuroessentialism, because it assumes emotion in humans and emotions in various other animals to be the same. It operates as a synecdoche, allowing one to stand in for the whole.

In Olds's experiments, pleasure becomes localized in specific brain regions and specific actions, namely, pressing the lever to activate that brain region. The fact that the rats pressed the lever repeatedly and feverishly was taken as an indication that the action must be a reward—that it must be pleasure—and that the pleasure must exist as a thing. Were the rats enjoying themselves—even though they sometimes pressed the lever so much that they forsook food and any other kind of reward? Olds does not define pleasure in terms other than by the actions performed. In the behaviorist tradition, he eschews emotional terms that cannot be directly observed in the rats' behavior. But what is lost by understanding pleasure that way? Together, these metaphors literalize pleasure as a binary and as equal to reward, thereby concretizing pleasure as a unified thing that can be located in a specific brain region.

This conception of emotion still undergirds much of the research conducted using newer methods such as fMRI. To ravel out, or puzzle through this research, consider how it is typically operationalized. Typically, researchers will measure BOLD activation in participants before and after a stimulus. In doing so, as the ethnographer Simon Cohn puts it, neuroscientists construct a "spatio-temporal line" that "serves to ensure the demonstration of the materiality of something as apparently immaterial as the experience of pleasure."[20] That spatio-temporal line constrains the relevant space to that of the brain (or even to particular regions of the brain) and to the temporality of the instant directly after a stimulus is provided in the scanner.[21]

---

19. Olds, "Self-Stimulation of the Brain," 314.

20. Cohn, "Petty Cash and the Neuroscientific Mapping of Pleasure," 152–53.

21. Note that in fMRI studies, researchers generally choose between conducting a scan of the whole brain or of a particular region or set of regions that they hypothesize may be relevant to the research question. The latter is known as "region of interest." See Jack and Appelbaum, "'This Is Your Brain on Rhetoric.'"

This assumption was not necessarily in place in earlier psychological traditions. Herbert Spencer, for instance, describes how a pleasurable stimulus, such as smelling a rose, diminishes in intensity over time, which can in fact lead not to pleasure but to disappointment. Emotions, Spencer suggests, have duration: "Even a simple emotion, as of anger or fear, does not reach its full strength the moment the cause presents itself; and after the cause is removed it takes some time to die away."[22] Emotions may come in "gusts or bursts" but may also last for hours or days, Spencer writes, but they aren't uniform: they entail a "succession of rises and falls of intensity," often interspersed with calm.[23] Unless specifically designed to do so, the typical fMRI experiment does not capture this temporality of emotion. The metaphor of the region or brain, constrained in space, is also therefore constrained in time; the assumption that neural stimulation is "all or nothing" vacates any since of duration.

As Henri Bergson might put it, neuroscientific measures of emotion render states of consciousness as material, assuming that they can be measured, counted, and spatialized as locations of action in the brain. Yet Bergson would distinguish between the material objects, "to which the conception of number is immediately applicable" and "the multiplicity of states of consciousness, which cannot be regarded as numerical without the help of some symbolical representation, in which a necessary element is space."[24] In social and cognitive neuroscience, these states of consciousness are symbolized mathematically and then translated into visual symbols. These symbols are taken to stand in for conscious states, which are spatialized and frozen in time, but, as Bergson might suggest, emotions are not experienced in these snapshots but diffused through duration, the pure experience of time. Yet in fMRI, the material constraints of the tool and its operation require experience to be spatialized and frozen in thin slices of time.

Further, as Cohn describes it, in fMRI-based cognitive neuroscience research seemingly "external" elements, such as "the dynamics between individuals, social expectations and the broader impact of culture" (which have both duration and space), are not considered relevant in this temporal-spatial pathway, so they are excised.[25] The body itself, and its history, are excised, despite Spencer's note that an emotion "habitually works changes, external and internal, throughout the body at large," which can be observed in terms of

---

22. Spencer, *The Principles of Psychology*, 121.
23. Spencer, 122.
24. Bergson *Time and Free Will*, 87.
25. Cohn, 153.

respiration, circulation, digestion, movement, and the like.[26] fMRI machines simply cannot take account of that embodiment on their own. (Some embodied measures, such as measures of skin conductance, heart rate, or respiration may be possible, but other measures, such as movement, would be forbidden due to the limitations of fMRI technology.) Yet those elements work to constitute emotion in an fMRI study, along with the fMRI machine itself, the stimuli used to induce pleasure, the participants' embodied experiences, the forms used to screen and consent participants, and so on.

Importantly, metaphors (as part of the rhetorical infrastructure of cognitive neuroscience) allow that substitution to take place: metaphors of networks, seats, centers, information, and so on *stand in for* this broader array of things and allow those other things to be ignored, becoming synecdoches for the broader assemblage of people, ideas, and things that are enacted when an fMRI experiment takes place. As Cohn puts it, this type of fMRI study enacts a version of materiality in which stimulus and response can be isolated and rendered knowable, located in a seat or as a function of a network. The metaphor of the "seat" or "center" transfers meaning in two ways, then: there's the linguistic transfer of meaning, from the idea of a "seat," say, as a regional or geographic center that is carried over to the "seat" of a particular neurological function in the brain, but there is also a material-semiotic transfer in which that metaphor is carried over into the experimental apparatus itself, as we see with the rat-treadle-brain circuit Olds generated that literalized the metaphor of the network. And yet there is one more transfer of meaning that occurs after the experiment, in which the material assemblage of the experimental apparatus is carried over and condensed into the metaphor of the "pleasure center" (an operation that we might describe as synecdoche). If figures can epitomize arguments, here they epitomize arguments that are material *and* semiotic.

Those metaphorical innovations that I have traced thus far help to make that isolation possible and are hence part of the rhetorical infrastructure of social and cognitive neuroscience, that which is necessary to produce knowledge in that field (at least as it is currently practiced using fMRI). In other words, metaphors are as much a part of the scientific apparatus as is the fMRI scanner or the screens onto which images and text are displayed or the brains of participants themselves. These metaphors form the basis of neurorealism, without which researchers could not claim to be identifying anything at all when they study someone in an fMRI machine. I do not mean to suggest here that researchers are necessarily to blame for taking their objects of study to

---

26. Spencer, 123.

be "uncritically real"; they are simply enacting events within the methods and traditions of a field that requires researchers to work on the assumption that those objects are real in order to be able to study them in the first place.

This assumption shapes the material practices of neuroscience in another way. As Cohn notes, researchers use statistical methods to determine what counts as activity in the brain: they must determine some threshold below which there is "no action" and above which there is "action," when really it is a matter of which brain regions are "more or less" active. Beyond this decision, they must then attempt to delineate regions of activity "as though they were definitive structures," in some cases even tracing perimeters of regions by hand in order to make them appear more concrete.[27] They can tinker with image settings that allow them to adjust the color and contrast—much like Instagram filters—and may select those filters that heighten the sense of concreteness of those brain regions. In this way, Cohn argues, "neural activity is converted not only into colourful objects, but material entities."[28] Thus, through a research method that is not just looking for but producing regions, the metaphor of the "seat" or "center" becomes literalized as a concrete site in the brain.

Notably, many scientists would agree that the "pleasure center" topos is outmoded and not reflective of scientific knowledge, and many are seeking to advance richer concepts of pleasure that take into account a broader range of factors. For instance, in a review in *Nature Neuroscience,* Samantha Leknes and Irene Tracey note that primary studies have identified no fewer than eleven major brain regions involved in pleasure. For instance, the lateral prefrontal cortex is linked with pleasurable tastes, while the anterior insula is linked with pleasurable responses to food, including chocolate, specifically. Notably, these regions overlap (several regions are implicated in responses to music and chocolate, for instance), *and* they are also implicated in responses to pain.[29] Clearly, pleasure responses are much more complex than Olds's rats would indicate. Interestingly, Leknes and Tracey cite both Plato and Aristotle in their review, noting that Aristotle's "Golden Mean" seems to have relevance for understanding the modulation of pain and pleasure in humans.

Thus, both the approach typically used in the "affective turn" in the humanities and the classical approach in neuroscience suffer from a series of limitations:

- They assume a primarily reactive (stimulus/response) model rather than a *predictive* model.

27. Fahnestock, *Rhetorical Figures in Science,* 160.
28. Fahnestock, *Rhetorical Figures in Science,* 162.
29. Leknes and Tracey, "A Common Neurobiology for Pain and Pleasure."

- They can overgeneralize/universalize, assuming that affective/emotional responses are evolutionarily determined, innate, and instinctive
- They can re-enact a mind/body (affect/emotion) dualism that fails to account for the experiential raveling of those elements
- They can tend toward naïve realism and essentialism
- They decouple discourse and affect

In the next section, I provide some suggestions for how researchers might address these limitations.

A NEURORHETORICAL APPROACH TO AFFECT

Both rhetoricians and neuroscientists have begun to articulate a different model of affect and emotion. By raveling together work in these fields, we can develop a richer theory that situates affective responses within sociorhetorical contexts without excluding its materiality. Contesting the classical model of affect/emotion, neuroscience researchers Lisa Feldman Barrett, Kristen Lindquist, and others have taken on the longstanding scientific paradigm that considered emotions to be pure, evolutionarily based responses to environmental stimuli in "reward systems" or "pleasure centers" of the brain. Instead, these researchers posit a constructed theory of emotion. But rhetorical scholarship would push this constructed view further, to consider its *situatedness* and its orientation toward objects, other people, and places; for these scholars, emotions are not only constructed but contextualized.

Kristen Lindquist, Maria Gendron, and Ajay B. Satpute describe their model of emotion as the psychological constructionist model. According to this theory, it is not just that words name emotional states, but that words *constitute* emotional states. That is, "language might not just translate feelings into words, but might help shape the nature of those feelings to begin with."[30] Lindquist, Gendron, and Satpute draw on a range of psychological studies to support this hypothesis. For instance, they note that children cannot reliably distinguish between emotions beyond "happy" and "sad" until about age four; prior to that, they generally lack the vocabulary for words such as *afraid* or *disgust*. Similarly, adults with cognitive impairments that limit their ability to process words also have difficulty processing emotions in studies that ask them to sort pictures of faces on cards into discrete piles. While unaffected individuals could sort cards into six different categories

---

30. Lindquist et al., "Language and Emotion," 2.

(happy, sad, disgusted, fearful, etc.), affected individuals usually sorted cards only into pleasant/unpleasant/neutral. Further, they cite linguistic and psychological studies that point to the variability of emotion words across cultures. Words for particular emotions known to English speakers may not necessarily exist in other languages. Meanwhile, English has borrowed words for emotions such as *Schadenfreude* (from German) to describe emotions that were not previously named in English. These and other instances lead Lindquist, Gendron, and Satpute to hypothesize that "emotions are experienced when affective states are made meaningful as specific instances of the emotion categories that exist in a given culture."[31]

Similarly, Lisa Feldman Barrett and Ajay B. Satpute argue for a "contextually sensitive, constructionist approach to understanding the neurobiological basis of emotion."[32] As justification for this approach, Barrett and Satpute cite meta-analysis studies that show that, contrary to a "classical model" that understands each emotion as a unique, "natural kind" with its own set of neurological responses—as in the notion of a "pleasure center"—neuroscience studies to date show considerable overlap between different emotions and the brain regions that correlate with them. That is, there is no discrete set of neurons, or regions, or networks for each emotion. In Barrett and Satpute's view, a theory of emotion should account for the following:

    1. The "recognition that definitions of emotion are stipulated, not discovered."[33] This proposition challenges researchers to recognize how their own suppositions about an emotion guide research studies. Much emotion research, Barrett argues, has first proceeded by using an assumed definition of a particular emotion. That definition then dictates the questions asked, experimental design, and interpretations offered. These emotions are always "recognizably English."[34] This approach "assumes that the emotion categories we experience and perceive as distinct must also be distinct in nature."[35] Thus, for instance, in Olds's experiments (which epitomize this "classical model"), pleasure is assumed to be a distinct, clearly delineated state that is separable from pain and has a simple on/off switch in a particular brain region, or "center."

---

31. Lindquist et al., 5.

32. Barrett and Satpute, "Historical Pitfalls and New Directions in the Neuroscience of Emotion," 5.

33. Barrett and Satpute, 5.

34. Barrett, "The Theory of Constructed Emotion," 1.

35. Barrett, "The Theory of Constructed Emotion," 1.

2. Emotions are not simply reactions but constructions of the world that involve meaning-making, and this is a predictive activity—it requires past experience or memory. However, this is seldom tested because current studies (and apparatus) assume a stimulus-response model. That is, according to Barrett, the brain isn't just passively waiting for a stimulus but is always engaging in activity, even without an external stimulus. In this view, the brain is always engaged in prediction of a "sensory array" and how it is actionable, meaningful, based on past experience.[36] Thus, any state that might be given the name "pleasure" is not purely a *response* to a stimulus but in part a prediction based on past experience. For Olds's rats, this is evident because they must first learn that pressing a lever produces a pleasurable response. The "reaction" from pressing the lever entails a prediction as much as a response. The anticipation of pleasure constitutes pleasure itself. (Consider how anticipation feeds into pleasurable activities on a cultural level: waiting in line for a concert, say, or—at least for those who enjoy theme parks—the feeling you get as a roller coaster climbs higher and higher right before plunging into a fall or barreling into a loop-de-loop.) Rather than understanding emotions as discrete or universal, Barrett suggests that they should be considered relative and situationally dependent: "A given emotion is a category of variable instances that are tuned to the situation at hand."[37] Ultimately, for Barrett, emotions should be understood as complex, dynamic interactions: "You construct emotions as experiences or perceptions—they emerge from complex dynamics within your nervous system, which is constantly in dynamic interaction with the surrounding context, often including other creatures who each have a nervous system."[38]

Barrett, Satpute, Gendron, and Lindquist (along with other scholars working in this area) challenge the model that undergirds Olds's studies and also the last few decades of neuroscience research. In this way, they understand emotion in ways that are more compatible with rhetorical theories and concepts.

Scholars familiar to rhetoricians have, or course, traditionally understood pleasure as multiple, timely, individual, and situation-dependent. Plato did not understand pleasure merely in what we might call the "classical" sense, as the sensation of return from a state of deficiency to a healthy state, or as the fulfillment of needs. Instead, Plato noted that one could also feel pleasure in sensations felt without a prior need, as in the pleasure one gets from learning,

36. Barrett, "The Theory of Constructed Emotion," 9.
37. Barrett, "The Theory of Constructed Emotion," 7.
38. Barrett, "Navigating the Science of Emotion," 43.

or smelling, or seeing. In the *Philebus*, Socrates proclaims that pleasure (personified in the goddess Aphrodite)

> I know to be manifold, and with her, as I was just now saying, we must begin, and consider what her nature is. She has one name, and therefore you would imagine that she is one; and yet surely she takes the most varied and even unlike forms. For do we not say that the intemperate has pleasure, and that the temperate has pleasure in his very temperance—that the fool is pleased when he is full of foolish fancies and hopes, and that the wise man has pleasure in his wisdom? and how foolish would any one be who affirmed that all these opposite pleasures are severally alike![39]

Throughout this dialogue, Socrates and his interlocutors establish that there are several types of pleasures: "pure pleasures," such as single tones, shapes, and the like; "mixed pleasures," such as the bittersweet pleasure of scratching an itch; and "pleasures of knowledge," which do not stem from a felt need in the same way as the pleasure that one gets from quenching one's thirst. Thus, for Plato, the idea of a single "pleasure center" in the brain would seem to contradict his view of pleasure as taking "the most varied and unlike forms."

In the *Nicomachean Ethics*, Aristotle similarly argues against a uniform view of pleasure as the replenishment of a need, since "the pleasures of learning and, among the sensuous pleasures, those of smell, and also many sounds and sights, and memories and hopes, do not presuppose pain."[40] Aristotle also notes that, in some situations, pleasures can decrease over time, since "at first the mind is in a state of stimulation and intensely active" but later grows "relaxed; for which reason the pleasure also is dulled."[41] For Aristotle, different kinds of pleasures are bound up with different kinds of activities. Further, pleasures carry values, since some are good and some are bad, in accordance with the activities they complete: "The pleasure proper to a worthy activity is good and that proper to an unworthy activity is bad." Aristotle insists that pleasures are unique to each species, since "horse, dog, and man have different pleasures."[42] He also argues that pleasures differ among individuals, "in the case of men at least," since "the same things delight some people and pain others, and are painful and odious to some, and pleasant to and liked by others."[43] Aristotle might question, first, whether the "pleasure center" in a rat could

---

39. Plato, *Three Dialogues: Protagoras, Philebus, and Gorgias*, 56.
40. Aristotle, *The Nicomachean Ethics*, 10.3.
41. Aristotle, *The Nicomachean Ethics*, 10.4.
42. Aristotle, *The Nicomachean Ethics*, 10.5.
43. Aristotle, *The Nicomachean Ethics*, 10.5.

correspond to that of a human, given that each animal has "different pleasures." And he might further argue that not all the rats in Olds's study might have delighted in pressing the lever—some may have been experiencing pleasure, he might argue, while others might have been experiencing a compulsion, a desire to alleviate boredom, or some other mental state.

From Plato and Aristotle, then, we would expect that pleasures are of different types and intensities, depending on their objects or activities, that they carry different values (some good, some bad), that they depend on the species in question, and that they depend on the individual's experience and taste. To call the rat's self-stimulation "pleasure," then, assumes that pleasure is a single thing that can be located in the brain. The metaphors driving neuroscientific inquiry ensure this singularity, insofar as they are based on the idea of a binary (on or off) function of the nervous system that they draw from cybernetics and computational metaphors and a singular center or seat (drawn from locationist metaphors).

Burke would also insist that pleasure be understood in relation to the orientation that an individual forms around an activity. As Debra Hawhee and I have (separately) argued elsewhere, Burke's research on drug addiction at the Bureau of Social Hygiene in the 1930s informed his concept of "piety" in *Permanence and Change*. Burke used the example of the drug addict to demonstrate how individuals build orientations around an activity according to a sense of the appropriate. For the drug addict, the physical pleasure (or affect) experienced from the drug becomes symbolically linked to other elements—it becomes part of the material, symbolic, and embodied cognitive system that Burke calls "orientation" or "piety."[44]

Thus far, however, the psychological constructionist theory of emotion seems to maintain a separation between affect, understood psychologically as the "basic feelings from the core of the body," and emotion, understood as the "psychological states that are experienced as coordinated patterns of physiology, behavior, and thoughts that occur within certain types of situations."[45] Yet, Lindquist, Gendron, and Satpute also present experimental results that suggest a more complicated relationship between affect and emotion. In particular, they suggest that "the accessibility of emotion concepts influences how feelings manifest as physiological responses during emotion"; namely, in experiments, researchers have found that labeling emotional states can lead to reduced embodied responses, such as heart rate and cardiac output (one measure of affective response).[46]

44. Hawhee, "Burke on Drugs"; Jack, "'The Piety of Degradation.'"
45. Lindquist et al., 4.
46. Lindquist et al., 11.

In one study, for instance, researchers asked a group of participants to complete a practice Graduate Entrance Exam (GRE) test. In the control condition, participants simply completed the test and then had their saliva sampled to test for serum amyloid A (sAA) levels, which indicate arousal of the sympathetic nervous system. In the test condition, participants completed the test after hearing a short explanation stating that, while it is common to feel anxious during a standardized test, anxiety could actually improve performance. These participants were instructed that if they were feeling anxious, they should remind themselves that anxiety could actually help them do well.[47] Those participants in the test condition not only experienced greater sympathetic nervous system activation; they also performed better on the quantitative portion of the GRE (the one that is, at least for many participants, the most nerve-wracking). Surely, limitations exist for this type of experiment, including small sample sizes and the question of whether the testing environment could fully capture the actual anxiety that GRE test-takers experience when entrance to graduate school is on the line. The implications of the study, however, are intriguing for an understanding of affect. The authors suggest that "the manner in which internal states are interpreted can have profound effects on emotions, physiology, and behavior."[48] In this case, then, we find a model in which affect does not simply precede or lead into emotion but in which affect and emotion feed into each other and, crucially for rhetoric scholars, are affected by language. These studies seem to suggest, then, that affect, emotion, and language are not easily dissociable practices but are instead raveled together.

Further studies suggest that sensation is also raveled into the language/affect/emotion complex. For instance, a body of research suggests that color perception is affected by linguistic labeling. Much of this research draws on cross-cultural comparisons, showing that in languages where more distinctions are made between, say, light and dark blue, participants recognize more gradations in that color. According to Gary Lupyan, "On this account, long-term experience categorizing the color spectrum using language gradually warps the perceptual representations of color resulting in more similar representations of colors in the same category (i.e., those labeled by a common term) and/or less similar representations of colors grouped into distinct categories (i.e., those labeled by distinct terms)."[49] Researchers also suggest that they can manipulate participants' perceptions in research studies by introducing them to new color terms and training them to recognize a color boundary

47. Jamieson et al., "Turning the Knots in Your Stomach into Bows," 209.

48. Jamieson et al., 211.

49. Lupyan, "Linguistically Modulated Perception and Cognition," 2.

that they did not formerly perceive, in a process called perceptual warping.[50] For Lupyan, these studies indicate that language may produce "transient modulation of ongoing perceptual (and higher-level) processing," such as processes of color identification and categorization.[51] Importantly, though, Lupyan recognizes that these linguistic effects do not happen all the time but only in specific situations and tasks in which language is evoked.

Although not a neuroscientist or psychologist by training, Margaret Wetherell offers an especially useful way of putting this work to use in the context of discourse studies. Drawing on work by Barrett and others, Wetherell argues that humanities scholars should adopt a theory of affect that recognizes its imbrication in language, culture, and habit, rather than one that posits affect to be somehow separable from or prior to those elements. Instead, she suggests, we should understood affect in terms of practices or "repertoires" that "emerge in bodies, in minds, in individual lives, in relationships, in communities, across generations, and in social formations."[52] Rather than splitting off affect from emotion or discourse, Wetherell encourages scholars to consider them as mutually constitutive. She cites, for instance, the "feeling bubbles" that emerge in national life around certain events, such as New Year's Eve or the anniversary of 9/11. These "feeling bubbles" represent cultural practices that effervesce through various modes and media, interpellating individuals into a collective affective bubbling.[53] They are also variable, however, and models of "affective communities" do not always account for this variability.

This conceptualization of affect, emotion, and language seems to align better with the theories of affective community or "structures of feeling" that humanities scholars like Sarah Ahmed, Lorraine Cvetkovitch, Lauren Berlant, Larry Grossberg, and Jenny Rice posit. A humanistic approach tends to add a deeper layer of understanding, a focus on sedimented histories, social practices, and discourses that might shape the individualized experience of affect or emotion privileged in psychological studies.

In this view, as Rice explains it, emotions are not best understood as possessions—"something we either have or don't have." Instead, following Ahmed, she argues that we should consider them as "affective orientation devices," ways of relating to others (including people as well as objects and places).[54] Rice describes how publics participating in debates about urban development often orient toward such debates through the topos of injury, an

---

50. Lupyan, 2.
51. Lupyan, 4.
52. Wetherell, "Trends in the Turn to Affect," 147.
53. Beer, "The Future of Affect Theory."
54. Rice, *Distant Publics*, 55.

affectively laden position that emerges when the terms of a debate are structured around how relevant parties have been (or may be) damaged by a course of action. The feeling of being injured is not simply a discursive framing, however, but one that invokes people's ties to places and objects, their relationships to other people, to histories of persecution and discrimination, and so on. For instance, Rice describes how people opposed to an urban development in Austin, Texas, put forward potential injuries to a local aquifer and swimming hole as part of their arguments against the project, but their connections to these locales were also consistently activated by participants in the debate through speeches, songs, and even short films designed to bolster opposition to the development. As Rice puts it, "affect is structured at multiple points of institutional, cultural, social, and educational sites."[55]

Daniel Gross and Stephanie D. Preston (a rhetorician and a neuroscientist, respectively) offer one example of how this rhetorical approach can address neuroscience research. For instance, they complicate neuroscience studies of race relations that suggest that white Americans unconsciously respond to black faces with fear (or, at least, with correlated blood-oxygenation-level-dependent [BOLD] activity in the amygdala). While neuroscience studies might interpret these results as vaguely reflective of a history of race relations in America, Gross and Preston argue, they do not go far enough to consider the social structures and histories that give rise to such a response (such as the legacy of slavery and Jim Crow laws). That is, according to Gross and Preston, "racism is not located primarily in the brain," as a response to a situation, but is built into situations.[56] Those best suited to analyze those situations—as historical, social practices—are humanists, but they can work with neuroscientists to develop richer types of experimental practices.

The richest rhetorical theories of emotion and affect, then, stress the importance of discourse as well as that of rhetorical situation while also including an understanding of sedimented sociohistorical practices. When done well, rhetorical analyses of affect also pay attention to the variability of affects invoked—including responses that may seem like apathy or boredom. This is important because, in some cases, an emphasis on collective structures of feelings can overemphasize that collectivity to the extent that a certain kind of "national" affect is posited, one that overlooks the variability among individuals in their orientations to a national event like 9/11.

Overall, this body of research posits that emotion, affect, language, thought, and sensation are not separable but are intertwined modes that acti-

---

55. Rice, *Distant Publics*, 97.

56. Gross and Preston, "Emotion Science and the Heart of a Two-Cultures Problem," 106.

vate (or are brought "online," to borrow a ascendant metaphor in neuroscience) in particular situations or accomplish particular tasks. (It goes without saying that using the term *online* to describe how certain cognitive systems or behaviors get put to use is clearly drawn from our understanding of computer networks and the internet.) These situations, in turn, do not simply appear out of nowhere but are built up out of cultural and historical traditions and institutions. Together, these constructed or constitutive views of cognition suggest that affect and emotion can be understood in the following terms:

1. As predictive or anticipatory, rather than merely a response to an external stimulus. That is, cognition "feeds forward" as well as backward.
2. As contextual and situational
3. As cultural and historical
4. Habituated or practiced, but not to the extent that habits cannot be shifted (experimentally, through "interference" tasks)

Building in those four elements within an fMRI type of study is difficult, but not impossible. For instance, in a study of altruism, researchers Helen Y. Weng et al. tested whether participants' neural responses to suffering would change after participating in compassion training. One might question their measure of altruism—a game that involves distributing money to victims in a simulated dictatorship. (For those not living in a dictatorship, the scenario seems a bit far-fetched.) However, this study contains the kernel of an interesting idea—that participants' emotional responses are subject to changes based on training. By extension, one might consider other kinds of "training" that are not necessarily so formalized, such as through exposure to articles, books, images, sounds, and the like. Numerous studies point to the fixity of fairly negative affective responses, such as negative emotional responses to faces of those considered outsiders or Others. A more compelling line of research might investigate whether, and under what conditions, people can be prompted to feel differently.

To conclude, let's return to the Tragically Hip tour with which I began my analysis. What is going on during those final concert performances? Those who attended the concerts, watched the live stream on TV, or watched videos of the performance after the fact all came to that experience already primed to respond in certain ways. It is likely that most viewers had at least some prior exposure to the Hip and their songs—and many were diehard fans. They also probably knew, as well, that Downie had been diagnosed with terminal brain cancer, since this fact was widely reported in the news and social media. And they likely knew that this was to be the final set of concerts for the band. Thus the concerts were already situated as epideictic occasions: a chance to both celebrate a favorite band and to mourn the impending passing of Downie.

To take one example, consider a video compilation that shows fans gathered around the country to watch the final encore in the final Hip performance.[57] We see a group of people gathered at the Commodore Ballroom in Vancouver to watch the concert unfold live on large television screens. The band then begins the first notes of their hit song, "Ahead by a Century." Some fans are jumping up and down, in particular an especially perky woman who is bouncing up and down, hands raised in the air; some of the more reserved people in the crowd simply shuffle side to side. The video cuts to Bolton Street in Bobcaygeon, Ontario, and then to the National Music Center in Calgary. We see groups of fans outside Springer Market Square in Kingston, Ontario, where the concert is happening, hands raised in the air. The mood is mostly jubilant as the video cuts to Grand Parade in Halifax, and fans start to sing along as Downie begins the opening line of the song "First thing we climbed a tree . . ." But then the video moves to the Horseshoe Tavern in Toronto, and we see a close-up of a man with his hand over his mouth, his eyes red-rimmed and teary. He wipes his eyes. Back in the Commodore Ballroom, bouncy woman is still bouncing, ecstatic, but a few frames later we see another woman singing along to the song while tears stream down her face. On Bolton Street, a woman dances with a Canadian flag. As we move from scene to scene, we see a range of facial expressions, bodily postures, sounds, gestures, and movements. The affective atmosphere cannot be characterized by its universality or particularity—only by its profuseness and range.

One way of reading this event would be to say that participants were responding to a stimulus: the music of the Hip, the movement and reactions of fellow fans, the discursive and cultural context in which they were embedded. But in another sense, this event is also marked as a scene of anticipation. People certainly have different reasons for attending a rock concert, but it is fair to say that most anticipate a certain kind of experience, whether that's hearing your favorite song played live, engaging in the collective experience of listening to your favorite band with a large crowd of fellow fans, or getting wasted in the parking lot beforehand. Consider the sense of anticipation that occurs toward the end of the concert, before the encore. Affect here is both reactive and anticipatory; the two loop back and forth. As it is currently theorized, *neither* neuroscientific nor rhetorical theories are very good at addressing the forward-feeding, anticipatory affect and its links to discourse, culture, and

---

57. Maclean's, "The Tragically Hip's Last Song Rings Out across Canada."

materiality, because they both generally remain mired in a stimulus-response model inherited from classical psychology.

In the case of the final Hip concert, it is clear that everyone in attendance carried with them a certain set of expectations and anticipations. For instance, fans knew they'd be experiencing a particularly patriotic and nationalistic form of collective feeling, since the Hip's songs often call on lesser-known moments in Canada's history and the band itself is known as "Canada's band." Before the show, Trudeau called the Hip an "essential part of what we are and who we are as a country," framing the event in patriotic terms.[58] The concert, and its broadcast across Canada on the CBC network, constituted an affective community or group that could feel together in overlapping—but not necessarily identical—ways. As Wetherell puts it, "We seem to be drawn to, empathize with, and are most likely to copy, imitate, and share the affect of those we affiliate and identify with, and those whom we recognize as authoritative and legitimate sources. Context, past and current practice, and complex acts of meaning-making and representation are involved in the spreading of affect, no matter how random and viral."[59]

As Papoulias and Callard recount, affect theorists sometimes hark back to a pure, unadulterated, arhetorical form of affect that is not beholden to discourse or to culture.[60] In the case of this event, though, untangling (unraveling) those threads seems difficult and, perhaps, would miss the point. The point is that those threads are tightly *raveled*—the affective mixes with the symbolic, the nationalistic, the discursive and cultural. Whatever neurological reactions people had in that setting (whether in the so-called pleasure center or a hypothetical "catharsis center") must have been intermingled with those other raveled threads. And it is precisely the tight raveling of those elements that makes an event like that last concert so deeply affecting.

This chapter has shown, I hope, not only that neuroscience can be strengthened if raveled together with humanistic insights but also that the opposite is also true. In this case, a deeper understanding of how neuroscientists are approaching affect might inform a richer concept of affect. Ironically, while humanists sometimes turn away from the cultural and symbolic in their quest for understanding affect, neuroscientists are themselves investigating a model that would account for the symbolic or rhetorical aspects of affect and emotion. In the conclusion, I suggest how we can build from the areas of overlap I've identified so far and what that might mean for humanities scholars interested in the neurosciences.

---

58. "Tragically Hip's Gord Downie."
59. Wetherell, 154.
60. Papoulias and Callard, "Biology's Gift," 31.

CONCLUSION

WHAT I HAVE ATTEMPTED to do throughout this book is, to borrow Latour's terms, to pry open the black box of social and cognitive neuroscientific research, to enter "through the back door of science in the making" to not only see how neuroscientific knowledge is made but to consider how it might be done differently.[1] I have tried to work both through "negative modalities" and "positive modalities" by identifying limitations and constraints in neuroscience research but also by identifying opportunities for alternative methods and interpretations that are already present within those fields. These glimmers of hope could be intensified with even more attention to humanistic insights, particularly the insights of rhetoric.

Throughout, I have focused here on the scientific experiment as a site of rhetorical inquiry. For the most part, scholars in the rhetoric of science have focused on texts—scientific articles, popular news reports, and so on—and on how rhetors use language to put forward truth claims, consolidate support for those claims, and so on. Increasingly, we have focused on visual and multimedia, and mathematical and statistical argument. Yet, the experiment itself remains understudied as a site of rhetorical production. Here, I have shown how the guiding assumptions and methods used in scientific experiments (in this case, in fMRI studies of the brain) create the conditions of possibility for rhetorical arguments. In the case of fMRI, the choice of study participants,

1. Latour, *Science in Action,* 4.

psychological tests, and experimental stimuli (along with statistical tests, as Christa Teston points out) are all relevant to the kinds of arguments that neuroscientists can make.[2] In this case, I have argued, those experimental elements all contribute to neurorealism and neuroessentialism—in fact, they are necessary for how neuroscience research using fMRI currently unfolds.

This is not to say, however, that such research couldn't be done differently. While I do not believe that humanistic perspectives can necessarily transcend these limitations, they might help us to work within them. In fact, I have tried to highlight possibilities for research that would still work within the constraints of fMRI that at least question assumptions about the hardwired nature of sexed/gendered behavior (chapter 4), political orientations (chapter 5), and affect (chapter 6).

Throughout, I have sought to model a heuristic that begins by "raveling back" the dominant metaphors, rhetorical figures, and definitions that shape research into a particular phenomenon. For instance, research into sex/gender and political orientations both function through antithesis, isolating key features that can supposedly divide humans into two discrete groups. Phenomena are often defined in ways that square with these key tropes; for instance, the figure of antithesis leads researchers to understand politics in terms of two discrete orientations, "left" or "right," which excludes the considerable range of political parties, opinions, and beliefs at play in any situation and overlooks the large group of people who may consider themselves "independents," not affiliated with any one political party.

After identifying key figures and definitions, one can then "ravel out," interrogating how a study has been set up by asking how the key concept is operationalized. As we have seen throughout this book, social and cognitive neuroscientists can operationalize a key concept using a range of tools, from psychometric tests and quizzes to specific types of prompts and games that users engage in within the fMRI (or before or after an fMRI scan). They can offer stimuli that are visual, verbal, aural, even olfactory. The choice of how to operationalize a concept is a rhetorical one, since researchers must persuasively show that they are measuring what they say they are measuring. For those of us who are seeking to understand neuroscientific experiments of this type, questioning how a concept is operationalized can help identify alternatives as well as limitations. We might also identify expertise that is lacking from a study team. For instance, in a study of how creative writers improvise, it might be useful to have a creative writing scholar on board who can provide a richer understanding of writing practices in their sociohistorical context.

---

2. Teston, "Rendering and Reifying Brain Sex."

Finally, in the "raveling together" move, one might bring interdisciplinary perspectives to bear on the design, analysis, and interpretation of fMRI-based research. I have suggested, throughout, examples of how experts in fields such as politics, sex and gender studies, visual culture, and the creative arts—as well as rhetoric—might contribute to neuroscience research by extending and contextualizing that work. While strategically adopting neuroscience findings can be productive, often this approach is hampered by limited engagement with neuroscience research methods themselves, which, I have argued here, play a significant role in shaping what counts as knowledge in that field, often by constraining how key concepts are understood so that they fit within the methodological framework of fMRI-based social and cognitive neuroscience. I have not extensively discussed statistical methods here, but Teston has shown that these, too, significantly constrain how social and cognitive neuroscience research can proceed. These methods have been subject to debate, as well, because they can sometimes inflate significance or produce spurious results.[3]

In closing, I would like to suggest how this method might prompt further inquiry for scholars in the rhetoric of science and technology. For one, further inquiry into the rhetoricity of experiments themselves will give us a deeper sense of how scientific rhetoric works. While I have focused primarily on scientific articles in this book, in doing so I have concentrated mainly on the methods sections, where researchers describe what they did and why. While methods sections are obviously condensed and (to some extent) sanitized narratives of research, they nonetheless illuminate how researchers define a hypothesis and determine how to study it. In the case of fMRI research, neuroscientists must define what it is they are studying (such as emotion, creativity, or political orientation) and then determine how to operationalize it within the context of an fMRI machine, using the assumptions of null-hypothesis statistical testing which requires them to differentiate their target quality from a resting state or a control. They make these choices not only based on feasibility but with an eye toward eventually convincing reviewers that their results are valid and persuasive. Thus, these decisions are fundamentally rhetorical. These same decisions ultimately delimit what neuroscientists mean when they talk about qualities of human behavior that are then often taken up in humanistic scholarship or in public discourse more broadly.

Of course, those interested in the rhetoric of scientific experiments have other methodological options, including ethnographic or participant-observer

---

3. Teston, "Rendering and Reifying Brain Sex." See also Margulies, "The Salmon of Doubt."

research, interview methods, and collaborative research with scientists. My own interest in neuroscientific rhetoric began as part of a collaboration with a group of neuroscience researchers who were themselves interested in how rhetoric works in their field. While I have not focused on this collaboration here, working with these researchers and listening to them talk about their research has profoundly shaped how I approached this book. One colleague, for instance, pointed out to me early on that it is important to pay attention to how researchers operationalize key psychological concepts, which led me to study emotion and reason and to compare the definitions used by neuroscientists with those that seem common in our field. (They didn't usually overlap.)

In that project, I worked with L. Gregory Appelbaum to articulate what a neurorhetorical approach might look like;[4] this collaboration later led to a metanalysis project that focuses on which psychological concepts have been most studied in social and cognitive neuroscience, and which brain regions they are connected to discursively in neuroscience article abstracts.[5] Over time, both of us began to understand how those psychological concepts and their definitions were rhetorical—that is, human behaviors come to be defined based in part on how they can be operationalized persuasively, so that definitions necessarily narrow or essentialize that which they purport to measure. Looking across multiple studies, we see that each of these concepts are themselves multiple; that is, there is no single "emotion" or "creativity" but instead multiple emotions, multiple creativities, as those concepts are defined, operationalized, and enacted through different experimental protocols. We might say the same of the brain itself.

My colleagues also pointed me to the fundamental question about the persuasiveness of neuroscience findings that prompted this book; namely the article by Racine, Bar-Ilan, and Illes that outlined the three claims (neurorealism, neuroessentialism, and neuropolicy) that they identified as problematic. I began to think about those claims, how they emerge, and why they circulate, and this prompted me to employ rhetorical theories of stasis and figuration to better understand them. I determined that these three claims operated at the stasis of fact, definition, and policy, respectively, and, as such, they represent a process, of sorts, for thinking through an issue.

Engaging with researchers gave me a deeper understanding of the knowledge game within which researchers worked. It is in this way, I believe, that rhetoric scholars will be able to move into what remains what Latour might call the positive modality. While it is possible to employ neuroscientific

---

4. Jack and Appelbaum, "'This Is Your Brain on Rhetoric.'"
5. Beam et al., "Mapping the Semantic Structure."

research in the mode of verification (taking neuroscience findings as facts to support our claims) or in the mode of provocation (taking neuroscience findings as tools for invention and analysis), we can also try to engage in the process of neuroscience research, which may lead to different insights and outcomes for neuroscientists and for us.

Thus, here I briefly outline a few of the findings from social and cognitive neuroscience that seem relatively stable, from a scientific perspective, and that seem to have bearing on rhetorical study. Of course, scientific consensus is always unstable and may shift over time, so whatever generalities I offer here should be understood as provisional, or as "stabilized-for-now." And one more caveat—a full review of neuroscientific concepts is beyond the purview of this short conclusion, so I leave it to other scholars to shift through the depths of neuroscientific knowledge (ideally, with a sense of its limitations and tendencies toward neurorealism, neuroessentialism, and neuropolicy fully in view). For this reason, I focus specifically on the areas of inquiry that I pursued in this book: pleasure, creativity, sex and gender, politics, and persuasion.

WHAT DO WE KNOW?

First, for neuroscience research to function at all, it depends on *brains (and bodies) that are affectable.* We may quibble about the conditions under which they are made to be affected, we may interrogate the tools and materials used to produce those effects/affects (from fMRI itself to statistical tools to various olfactory, visual, or auditory stimuli), but the fact does remain (I think) that *something* happens (by chance or by design) in an fMRI study. The brain's material is such that it can be stimulated (in a laboratory setting) but also that it can be changed. That *something* includes response to the study conditions themselves—not just the stimulus designed but the entire setup—the people, sensations, beliefs, and so on brought into play within that study. In other words, we can't divorce the material responses in the brain that are identified in fMRI studies from the rhetorical situation and context. It is also important to consider, however, the extent to which the brain may function in a predicting, orienting model and not just a response mode. As I suggested in chapter 6, current models of affect draw on this understanding of the brain as predictive, but so do models of the "embodied mind" or "extended mind" put forward by scholars like Andy Clark and Alva Noë.

Beyond this insight, neuroscientists have built up, over time, understanding of how particular regions of the brain correlate with specific topics, and they can be studied through meta-analysis visualization tools such as Neurosynth, which aggregates findings from over 11,000 articles to show users regions in the brain where key terms (such as *attention* or *decision-making*) have correlated in fMRI-based studies. Interpreting these results, though, poses a challenge and is fraught with potential misinterpretations, as neuroscientists are well aware. For one, as I have shown here, a psychological concept or term, such as *decision-making*, may have been operationalized in any number of ways in each fMRI study. In most cases, there is no standard approach for what accounts as "decision-making" (or "reason" or "humor" or "emotion"), so it may not be the case that, when those results are grouped in meta-analyses, one is actually comparing apples with apples. Instead, one must be persuaded that the wide variety of stimuli used to evoke "decision-making" in each fMRI study are measuring what is essentially the same thing—in short, one must buy into neuroessentialism and overlook the differences among scientific experiments.

Moreover, fMRI equipment varies: not only are there different makes and models of scanners but also researchers can choose different settings. All these details are included in scientific articles. Take, for instance, this information from an fMRI study of political judgment:

> The whole study was conducted on a 3-T Siemens Magnetom Trio wholebody MRI scanner in the Biomedical Imaging Technology Center at Emory Hospital. Brain imaging involved the acquisition of 30 axial slices of 3 mm thickness, acquired parallel to the AC-PC line with a matrix size of 64 × 64 over a field of view of 22 × 22 cm. Blood oxygenation level dependent (BOLD) contrast images were acquired (TE of 30 msec) using T2*-weighted gradient-echo, echo-planar pulse sequences with a TR of 2.5 sec for a total of 477 scans. In addition, a 3-D MP-RAGE sequence was collected at an isotropic resolution of 1 × 1 × 1 mm for 3-D anatomic analysis and visualization of task-related activations.[6]

Here, the authors are describing a series of choices made to record BOLD activity in the brains of participants. All these details, from the make and model of the fMRI machine to the timing of pulses to the resolution used, are presumably relevant. With different settings, different results might have occurred. Similar choices govern how the fMRI data are then processed and analyzed statistically. Here again, different choices could lead to different results. However, in Neurosynth, those methodological differences are erased.

---

6. Westen et al., "Neural Bases of Motivated Reasoning," 1949.

Thus, Neurosynth should be understood as showing not how topics or concepts really are located in the brain but how they have come to be associated with brain regions through layers of scientific articles, each based on different experiments performed at different places using different protocols, equipment, and settings.

For rhetoricians, one way of approaching these various mappings is as a topology. As I have argued elsewhere with my colleagues, these mappings can be understood as representing not how the brain is organized, per se, but as how neuroscientists talk about the brain, its organization, and its functions.[7] By examining these topologies we can identify rhetorical topoi that represent opportunities for further research—such as isolated sets of topics and brain regions that seem understudied—as well as areas that could be further integrated into the discipline as a whole. From a rhetorical perspective, then, these topologies offer insights into how a scientific discipline communicates at a macro level, a task that can then be deepened through more traditional methods of rhetorical analysis.

What continues to challenge me, as a rhetorician, though, is the question of whether humanists should care about the specific activations that scientists identify through their carefully controlled experiments. In other words, if I say to you that x stimulus reveals activation in the prefrontal cortex or amygdala, is that statement meaningful to you as someone who is not a neuroscientist? Why do we care? Do we care simply *that* correlations exist between an experimental stimulus and a response in a brain region? Do we care *where* it occurs? Why? What investment do we have in the particular correlations that neuroscientists identify in particular networks/regions, versus the simple "fact" that some type of activation occurs (within an experimental setting)? What do we learn about the concepts we are interested in? And what kinds of knowledge do we need to have about neuroscience in order to find out? This is where further research might offer a richer claim for the significance of neuroscience research for rhetoric, in my opinion, and for the humanities more broadly. In short, are we stuck at the fact stasis (does it exist?) and not sure where to go next.

A second finding that scientists largely support at the current moment is that *brains are changeable,* or plastic. These changes can be examined on a small scale in an fMRI study, but the discrete stimuli in an fMRI experiment pale compared with the constant bombardment of stimuli in daily life, or with the careful, repeated, embodied practices that lead to neurological change over time. Our brains are always changing, yet, paradoxically, it seems

7. Jack et al., "Mapping Rhetorical Topologies in Cognitive Neuroscience."

clear that continued use also leads to engrained patterns of activations, to the strengthening of certain neural networks at the expense of others. The tensions between permanence and change (to borrow Burke's title) could be interesting for rhetoricians to explore and might complicate our own theories about how rhetoric functions. Here again, though, the risk for humanists seeking to work with neuroscientific ideas is that we may rest at the stasis of fact. Simply pointing out that our brains are plastic and can be changed by X activity is a relatively uninformative statement. How are they changed by this activity? Are those changes desirable or not? What should we do to enact these kinds of changes (or not)? The latter are fundamentally humanistic questions to which rhetorical scholars might attend.

Third, while neuroscientists must necessarily constrain the bodies of their participants in order to work within the limitations of their tools (from statistical tools to imaging technologies like fMRI), *cognition seems to continually exceed the brain itself.* Consider, for instance, rapper Open Mike Eagle, who struggled with the confinement of the fMRI machine as his body sought to act as metronome while he rapped (chapter 2). We can also imagine the embodied, emplaced nature of sexual desire (chapter 4) or political beliefs (chapter 5). Consider also how listening to campaign messaging might differ at a march or rally, where multiple rhetorical elements converge (from signs to songs to speeches), where the body is fully engaged in standing, sitting, clapping, or moving, and where even the spatial organization of the environment may elicit political responses.

For the most part, though, these kinds of considerations are extracted out in an fMRI study, at least in part because of methodological constraints. (The recent development of "portable" fMRI tools may offer different options.) Notable, too, are the embodied, spatial, and cognitive elements of *participating in an fMRI study* as well as many elements of the experiment itself. Of course, this is generally true of all experiments. As Latour and Woolgar put it in *Laboratory Life,* once a scientific claim has been generated based on an experiment, "it is easy to forget that the construction of the paper depended on material factors. The bench space will be forgotten, and the existence of laboratories will fade from consideration. Instead, 'ideas,' 'theories,' and 'reasons' will take their place."[8] This is particularly important to keep sight of in neuroscience studies. Consider how Cohn describes his experience in his ethnographic study of fMRI, "Petty Cash":

---

8. Latour and Woolgar, *Laboratory Life,* 69.

The volunteer lies on the trolley and is wheeled into the large aperture. The researcher joins her team in a separate room behind a glass screen, amidst banks of monitors and electronic equipment. From this point on the experiment is symbolically initiated and a key division is established. The sense of separating off the individual is endorsed by the very physicality of the laboratory—once in the scanner there is usually only a tiny line-of-sight that might betray the existence of the researchers. Stimuli, usually images, designed to activate either positive or negative reactions, are directed to a screen above the volunteer's head. The newly created distance between the person in the scanner and the researcher serves to ensure that the emotional experience is confirmed as existing solely inside the volunteer's head. Meanwhile, the scientists stare at monitors, make adjustments, chat and joke with each other, and somehow deny the presence of the person as anything more than a focus for the investigation . . . Sometimes they communicate instructions or reassurances to the person through an intercom, but more often the only potential contact is a panic button that the volunteer clasps in their hand. Overall, the idea that the feelings derived might be shared or intersubjective is apparently circumvented through the very indirect delivery of disembodied stimuli.[9]

If they are acknowledged at all, these experiences are extracted from the experiment as it is described in research articles. While some hold out great hope for neuroscience and its ability to enrich our understanding of embodiment, as Cohn shows, fMRI research, at least, is strangely disembodied. To date, those who want to push for a richer understanding of cognition as embodied and enacted within an environment do not seem to have much influence over this particular area of research, for the reasons having to do with its rhetorical infrastructure that I have examined in this book. In addition, the statistical and empirical requirements of fMRI-type research currently push researchers to dichotomize, to separate a behavior from its opposite or from a resting state, in order to create a testable hypothesis with a control condition that is opposed to a test condition. This, too, requires extraction rather than immersion, since in real-time and real-place interactions, the mind and body continually move between, in, and through different kinds of behaviors and tasks.

To counter these limitations, some researchers (including many in rhetorical studies) express enthusiasm for what is called the "extended mind" theory

---

9. Cohn, "Petty Cash," 156.

or enaction theory.[10] Understanding the brain as embodied and emplaced provides an opening for the kinds of humanistic insights that I have outlined here; often, my analysis of fMRI-based scientific experiments and articles has pointed out ways that they lack that perspective. However, the impact of embodied cognition on fMRI-based neuroscience experiments is questionable at this point. In an interview with David Gruber, for instance, Jack Gallant (a professor of neuroscience at the University of California, Berkeley) states outright that "in large part, I've ignored the embodied cognition movement."[11] For Gallant, theories of embodied cognition, which stress the situatedness of the brain in a body that moves through an environment, are not sufficient to *explain cognition,* which is his goal as a neuroscientist. Similarly, when I asked a neuroscientist colleague what he thought of the work of Andy Clark—perhaps the person most commonly cited in relation to theories of embodied mind—he stated, "I know next to nothing about him. My guess is that the impact to neuroscience is minimal."[12] These could, of course, by idiosyncratic responses, but the ongoing scholarly debates about embodied cognition (EC) and its relationship to cognitivism suggest that the former remains controversial, at minimum. To date, scholars continue to debate how key terms in the "enacted cognition" or "embodied cognition" school should be defined and how research should proceed, including questions about what kinds of testable hypothesis it can produce, what methods should be used, and what kinds of evidence it can generate.[13] While I have focused on social and cognitive neuroscience studies that draw on a cognitivist paradigm in this book, it would be interesting to study the rhetoric of EC itself, as an incipient theory and method, to ask how researchers advance answers to those questions. For scholars in rhetoric seeking to draw on the insights of EC, additional questions emerge. Aside from the general notion that brain, body, and environment all participate in cognition, what specific implications of this would be particularly relevant for rhetoric scholars? What kinds of cognitive theories and methods would we need to develop in order to develop more nuanced theories of rhetorical cognition? How might EC allow us to study the interaction of brain, body, and environment in ways that add to what we already know about rhetoric?

---

10. See, for instance, Rickert, *Ambient Rhetoric*; Rivers, "Future Convergences."

11. Qtd. in Gruber, "NeuroHumanities."

12. Huettel, personal correspondence.

13. See, for instance, van Dijk et al., "Can There Be Such a Thing as Embodied Embedded Cognitive Neuroscience?" and Goldman and de Vignemont, "Is Social Cognition Embodied?"

WHAT WOULD WE LIKE TO KNOW?

What conditions, then, would make it possible for rhetoric scholars to gain new insights from neuroscience without losing what we value about rhetorical scholarship—context, embodiment, situationality, contingency?

1. We need a deeper knowledge of neuroscience itself. As I've suggested above, the kinds of research I've examined here are interested especially in the question of how behavior relates to brain activity, whether understood as localized in particular regions or as extended across different networks. If we are careful to situate these findings with respect to a sense of the embodied, enacted brain (cognition in the wild, so to speak) and are careful to understand how behaviors are being defined and operationalized, we still need to know more about brain anatomy and cognition in order to interpret those findings in ways that can extend our existing theories. Opportunities exist for humanistic scholars to gain those insights, including by collaborating with researchers and by attending workshops and training events such as the University of Pennsylvania's Center for Neuroscience and Society's Neuroscience Boot Camp.

2. By the same token, neuroscience researchers could benefit from deeper training in the humanities; we need the reverse of the Neuroscience Boot Camp—we need "Humanities Boot Camps" for neuroscientists. Humanists have thought deeply about many of the concepts that social and cognitive researchers think about. As Gabriel Abend puts it, "in order to construct their object, moral neuroscientists should draw on moral philosophy and the history, anthropology, and sociology of morals. Neuroscientists of spirituality and religious belief and experience should draw on theology and the philosophy of religion. And so on." It is not just that the humanities offer richer perspectives on these topics; Abend argues that in these disciplines researchers have thought carefully about the demarcations that neuroscientists often seek to make between related terms. Thus, in humanities scholarship we find "the most sophisticated arguments and counterarguments we have regarding the boundaries between, say, a moral norm and a convention or etiquette norm; or between loving someone, liking them a lot, and just being infatuated with them."[14]

Too often, because of our disciplinary silos and the kinds of academic training that undergraduates receive, scientists lack exposure to humanistic

---

14. Abend, "What Are Neural Correlates Neural Correlates Of?" 2.

scholarship. Similar divisions are being overcome in the field of medicine, where courses and programs in medical humanities abound. To a smaller extent, projects such as the Neurohumanities Research Group at Duke University[15] or Penn State's Neurohumanities Initiative[16] offer these kinds of opportunities for researchers to work across the traditional humanities/science divide. Further, changes in undergraduate training (including an emphasis on "STEAM") offer further impetus for integration of insights. It is notable that Kristen Lindquist, whom I featured in chapter 6 in relation to her constructed theory of emotion, was an English major as an undergraduate and explicitly seeks to integrate humanistic theories in her research. Currently, neuroscience researchers may see little reason to collaborate with humanities scholars or to recommend humanities courses to their students. In his interviews with prominent neuroscientists, Gruber asked participants to reflect on the disciplinary hurdles they face in collaborations with neuroscientists. Their responses were telling: some identified problems with locating funding or developing testable hypotheses, and others admitted to outright indifference to humanities scholarship that "doesn't really matter to me."[17] Thus, "academic cross-training" can be difficult, for students and researchers alike, but it offers opportunities for growth in both areas that can move beyond small-scale borrowings and flower into new research areas.

3. To accomplish interdisciplinary research, we need to develop a vocabulary that will allow researchers from the humanities and the neurosciences to work with each other. For humanists, this means that we should be able to explain our own theories and concepts in ways that nonspecialists can understand and that can relate to their own theories and terms. Admittedly, this is not an easy task, and many have tried to address the problem of "incommensurability," or the perceived difficulties in communication between different theoretical and methodological programs.[18] In my own collaborations, for instance, I was surprised that a member of our research team assumed that, by rhetoric, I was referring to "bias." We began one of our projects with a question about how rhetoric was working in research abstracts in neuroscience. I assumed that, of course, rhetoric is always operant in abstracts because any attempt to summarize research findings is necessarily a persuasive act. My colleague assumed that rhetoric could help us to identify "bias" in the abstracts, with the assumption being that bias could somehow be avoided. We

---

15. https://dibs.duke.edu/dibs/research/neurohumanities-research-group.
16. https://sites.psu.edu/neurohumanities/about-neurohumanities-initiative/.
17. Gruber, "NeuroHumanities."
18. Harris, "Introduction."

were using two different terms, rhetoric and bias, and we each had our own set of assumptions and theories in mind. Gradually, I was able to explain to my research group that, in my view, rhetoric was not simply a matter of bias. Terms such as *reason, affect,* and *emotion* are likely to pose similar challenges for researchers working in interdisciplinary teams, since the bodies of scholarship behind those concepts may vary considerably.

Two of us found more common ground working on an article as a pair ("'This Is Your Brain on Rhetoric'"). In particular, after the work I did to examine how terms such as *reason* and *emotion* were being defined and operationalized in a sample of articles, my colleague, Greg, began to say things like "of course, psychological concepts are rhetorical constructs." Of course, this kind of cross-disciplinary vocabulary building works both ways. In my case, Greg taught me the word *operationalize,* and it opened up a new line of inquiry for me, one that shaped much of the work in this book. These are all examples of terms common in one discipline and not another, but we also found ourselves inventing terms as we consolidated a research project that we conducted using social network analysis to study a corpus of research abstracts. For instance, in Beam et al. we used the metaphors of "islands," "bridges," and "gaps" to identify research clusters that were not linked to the general body of cognitive neuroscience research.[19]

4. We need to get over the kneejerk frames of "acceptance and rejection" that too often shape our approach to the sciences (and to neuroscience itself). As I have been working on this book, giving talks, and organizing workshops around this topic, I have often been puzzled when respondents assume that my goal is to "debunk" neuroscience or conclude that neuroscience is "junk science." This is not my point at all. As I see it, social and cognitive neuroscience (at least, what I have reviewed here) is working within a particular knowledge game to advance claims about the brain. Those claims tend to do with how behaviors or psychological concepts correlate with brain function and anatomy. Within that game, researchers use specific tools, including apparatus such as fMRI, psychometric scales, statistical analysis techniques and software packages, and visualizations, to put forward those claims. My goal here has been to show how, within that knowledge game, there are limitations that might be addressed through humanistic inquiry. Too often people seem to assume that I'm simply working in the "rejection frame."

On the flip side, however, I worry that the "acceptance" frame can miss some of the (rather important) limitations that I have outlined above, espe-

---

19. Beam et al., 1958.

cially when neuroscience is used to verify claims that scholars want to put forward (x is supported by neuroscience)—a phenomenon Jenell Johnson and Melissa Littlefield refer to as "fact finding." For Johnson and Littlefield, fact finding occurs when researchers use neuroscience findings (often from popular books) as "the ultimate arbiter of debates, as the objective voice in subjective matters."[20] This tendency is problematic, they note, given the rhetorical moves that writers make in popular books—often omitting hedges (the "mights" and "may suggests" of scientific language), drawing on appeals to wonder and application, and offering theoretical speculations that would not be found in scientific articles.[21] Johnson and Littlefield go on to note that neuroscience is also used in the mode of theory-building, which uses neuroscience (again, often through popular books) in much the same way that scholars might use critical theory or philosophy texts. Here again, the limitations of popular neuroscience books should be held in mind, Johnson and Littlefield warn, given their "tendencies toward neuroessentialism."[22] Johnson and Littlefield seem to suggest that these tendencies may be "balanced" by referring to the peer-reviewed scientific literature, suggesting that the latter can help to avoid neuroessentialism. My point throughout this book, however, has been that the peer-reviewed literature—and the experiments on which it rests—is also grounded in neurorealism and neuroessentialism as a condition of its production. Swapping out citations of popular books for citations of scientific articles does not necessarily mitigate against this tendency in scholarship that borrows from the neurodisciplines. Instead, researchers should pry open the black box, to use Latour's terminology, of the neuroscience experiment to better understand how neuroscientific knowledge claims are being produced, how concepts relevant to humanists are being defined and operationalized in scientific experiments, and with what limitations. Like Johnson and Littlefield, I am not arguing against attempts to perform transdisciplinary and interdisciplinary work between the humanities and neurosciences, only that this kind of work should involve deep engagement with the methods, theories, and techniques of all disciplines involved—including acknowledgment of ongoing debates[23] and possible limitations as well as their potential contributions.

It is important to recognize, ultimately, that sometimes the neuroscientific theories that seem most attractive to us as humanists may not be the

---

20. Johnson and Littlefield, "Lost and Found in Translation," 290.

21. Johnson and Littlefield, 286.

22. Johnson and Littlefield, 293.

23. As David Gruber notes in "The Extent of Engagement, the Means of Invention," scientific debates about neuroscience concepts are seldom acknowledged when those concepts are activated in humanities and social science research, at least in the case of mirror neurons.

same theories for neuroscientists, even when they share a name. The brain is, like the body, subject to "multiple ontologies" depending on where it is encountered and enacted. As David Gruber has shown, for instance, scholars in various disciplines have taken up the concept of "mirror neurons" in quite different ways and for different reasons.[24] Gruber argues further that humanities scholars "need not be content with the critique or the adoption of neuroscience as two exclusive moves, but might also harness strange, lingering, fantastic, familiar, or long forgotten epistemic resources to propose or enact other ontologies."[25] This inventional mode for neurorhetorics, Gruber suggests, is grounded in the idea of multiple ontologies, or a refusal to hand the brain over to neuroscientists as the sole epistemic arbiters of knowledge. This is all fine, in my opinion, as long as we don't lose sight of this multiplicity by positing universals where multiplicities exist.

How, then, is the brain enacted in rhetorical studies, and how do we want it to be enacted when we draw on neuroscience findings? A set of questions that Rose asked in 1988 about then current cognitive science findings and their relevance for writing studies still pertain today. Rose argued that the following questions should be asked prior to applying a cognitive science theory to the field of writing studies:

> Is the theory formulated in a way that allows application to writing; that is, can it be defined in terms of discourse? Given what we know about writing, how would the theory be expected to manifest itself? . . . What will the theory allow us to explain about writing that we haven't explained before? What will it allow us to do pedagogically that we weren't able to do as well before? Will the theory strip and narrow experience and cognition, or does it promise to open up the histories of students' involvement with writing?[26]

My focus in this book has been on rhetoric more so than writing studies, but these questions might be rephrased in terms of rhetorical theory:

- Is the theory formulated in a way that allows application to rhetoric; that is, can it be defined in terms of material-discursive practices and effects?
- Given what we know about rhetoric, how would the theory be expected to manifest itself?
- What will the theory allow us to explain about rhetoric that we haven't explained before?

---

24. See Gruber, "The Neuroscience of Rhetoric."
25. Gruber, "Reinventing the Brain, Revising Neurorhetorics," 250.
26. Rose, "Narrowing the Mind and Page," 296.

- What will it allow us to do pedagogically that we weren't able to do as well before?
- Will the theory strip and narrow experience, situational elements, historical and cultural factors, etc., from our understanding of cognition?
- Is the theory sufficiently rich to account for the variety of rhetorical practices, traditions, and effects?

The strategy of integrating neuroscience into rhetorical theory, I would argue, should consider questions. This is not to say that such applications are ill-advised, but just that they should allow for thoughtful answers to the above questions.

I would also argue, however, that we need not always be beholden to extant, published neuroscience research in our attempts to understand rhetorical cognition. Scholars in rhetoric and composition have a rich history of their own to draw from. As Nathaniel Rivers notes, for instance, researchers in rhetoric and composition already employ a robust set of research methods that might be valuable for cognitive inquiry, such as "videotaped writing sessions," "textual analysis of written artifacts produced with pen and paper and with computers," interviews with writers and "longitudinal participant-observer studies," and "genre tracing."[27] We might also add the methods of user-centered design advanced by Bob Johnson, chronotopic lamination used by Paul Prior and Jodi Shipka to examine writing processes, talkaloud protocols, and eye-tracking studies (all existing methods in rhetorical scholarship) as understandable as a form of cognitive inquiry in rhetoric.[28] Scholarship in technical communication and composition, I would argue, are especially rich given the range of research methods employed.

We can also engage in our own practical experiments, to engage in an "art of inquiry" based on the status of rhetoric as a *techne*. An art of inquiry, Ingold argues, is suitable for his home discipline, anthropology, which he aligns with art, architecture, and archaeology as disciplines that focus on human making. In an art of inquiry,

the conduct of thought goes along with, and continually answers to, the fluxes and flows of the materials with which we work. These materials think in us, as we think through them. Here, every work is an experiment: not in the natural scientific sense of testing a preconceived hypothesis, or of engineering a confrontation between ideas "in the head" and facts "on the

---

27. Rivers, "Future Convergences," 418–19.
28. Johnson, *User-Centered Technology*; Prior and Shipka, "Chronotopic Lamination."

ground," but in the sense of prising an opening and following where it leads. You try things out and see what happens.[29]

I would argue that rhetoric, as another *techne,* would also benefit from this type of inquiry, particularly as we seek to develop a rhetorical theory of cognition.

As a *techne,* craft, or practice, rhetoric has always been grounded in the contingent and emergent. Thus, our theories of mind and brain should hold that value in mind. As John Poulakos notes in his description of sophistic rhetoric, rhetoric, as an art, does not strive for "cognitive certitude, the affirmation of logic, or the articulation of universals"[30]—all of which are goals of fMRI-based cognitive neuroscience. Instead, Poulakos argues, rhetoric is oriented toward the possible: "the possible is not a given which can be known or verified; it exists in the future as something incomplete and dormant, something awaiting the proper conditions to be realized."[31] What, then, can we learn about rhetorical cognition, of possibility or *to dynaton,* through doing and making? Along somewhat similar lines, Thomas Farrell defines "rhetorical cognition" as involving "a quest for *meaning*" as opposed to a search for truth, with the goal being "to particularize meaning by instantiating and refiguring possible categories and criteria through the world of action."[32]

Scholars in indigenous rhetorics are laying the groundwork for this kind of inquiry. Qwo-Li Driskill, for instance, has described the embodied and cognitive actions at stake when learning basket-weaving: "my body literally had to learn how to make the movements to weave the basket. I had to learn how to hold my hands, how to hold my body, how to hold the basket reed, and suddenly something 'clicked.' And it wasn't verbal—it was kinesthetic."[33] To create meaning through basket-weaving involves these kinds of kinesthetic, embodied experiences that, Driskill finds, naturally blend with storytelling, another form of meaning-making. As Malea Powell argues, crafted items are *themselves stories* and need not be read as texts to have meaning. A basket, for instance "is MADE of story, it IS story";[34] as things, such artifacts of indigenous American tradition "provoke, create, and prompt the stories that tell us who we are in relation to one another." To understand these "things," Powell

---

29. Ingold, *Making,* 7.
30. Poulakos, "Toward a Sophistic Definition of Rhetoric," 37.
31. Poulakos, 44–45.
32. Farrell, "Practicing the Arts of Rhetoric," 195.
33. Driskill, "Decolonial Skillshares," 72.
34. Powell, "Rhetorical Powwows," 10.

argues, "we have to engage in the same kind of embodied, committed participation modelled by these makers."[35]

Making offers a form of rhetorical inquiry quite different from that of an fMRI experiment, one in which cognition extends across mind, body, material, and environment. Recently, I undertook to learn weaving. Learning weaving, I found myself, much like Driskill, focusing on the embodied actions that I needed to perform, copying those of my instructor and often asking her to show me several times before I got it. For example, to measure out the vertical threads (the warp) for my first project, we used a warping board, which is a board hung on the wall with pegs on it. To measure the warp, you first create a guide string the desired length of your warp, and then you figure out a pattern on the warping board that will allow you to wrap the warp thread around and around, so that each pass constitutes one warp thread. To create 100 warp threads, you must repeat this action 100 times, stopping periodically to count the number of passes so you don't get lost. Eventually, I got into a rhythm, passing the thread from right to left and left to right repeatedly, then stepping on a little stool to count the number of threads using a technique my instructor showed me (a motion that reminded me of cat's cradle games I played as a child). As Ingold and others have suggested, the kind of cognitive-embodied action used in crafts are forward-directed as well as feedback-directed,[36] and I definitely experienced this in the act of laying out my warp threads—I was both responding to the thread as it passed through my hands, sometimes correcting how I was holding the thread or the order in which I passed the threads around the pegs, but I was also anticipating where the thread would go next.

To take another example, the type of loom I used for my lessons is a jack loom. To operate this loom, you use your feet to press down floor pedals (much like the pedals on a piano or organ). Each pedal lifts a different set of heddles (metal "eyes" through which the warp threads are threaded); by lifting different sets of heddles you can create different woven patterns. For each weaving pattern, then, you lift the harnesses in a different order by pressing on the treadles in different patterns—such as "1 2, 1 2" (for a simple woven "tabby") or "123432, 123432" (for a twill). Thus, before I began each new pattern in my sampler, my instructor and I would practice the foot patterns on the floor, miming the direction in which I would be throwing the shuttle at the same time. Then, as I moved to the loom, I would repeat these patterns mentally and bodily, counting as my body moved in sync. My actions were

---

35. Powell, 15.
36. Ingold, *Being Alive*, 216.

directed toward what came next; I was "thinking ahead" rather than simply responding (although the loom, the shuttle, the thread, and the emerging pattern of woven fabric all provided "feedback" for my actions).

Rhetoric, I would argue, is similarly oriented toward both feedback and forwardness, and it might be useful to think about those loops alongside other *techne*. What would it mean to consider the rhetor, like the weaver, as, to borrow Ingold's terms, "not so much imposing form on matter as bringing together diverse materials and combining or redirecting their flow in anticipation of what might emerge?"[37] This view brings us, I think, to a renewed interest in *to dynaton* (or what John Poulakos defines as "the possible") as a direction for cognitive rhetorical inquiry through practical experimentation.[38]

We have, I think, focused on rhetoric in terms of response (both in traditional concepts such as Bitzer's rhetorical situation, in which rhetors respond to an identifiable exigence with rhetoric, and in more embodied, affective models such as Hawhee's and Rice's, in which those responses are not purely cogitative but embodied and affective). fMRI studies, too, tend to be response-based—that is, they rely on the classic behaviorist model of stimulus and response. What would it mean, then, to focus more on "forwardness"? on the thread that moves forward in place and time? For now, however, I lay that warp on the loom, so to speak, to be taken up and woven at another time.

37. Ingold, "The Textility of Making," 94.
38. Poulakos, "Toward a Sophistic Definition of Rhetoric," 36.

BIBLIOGRAPHY

Abend, Gabriel. "What Are Neural Correlates Neural Correlates Of?" *BioSocieties* 12, no. 3 (2016): 1–24.

"Air Travel in West Increases Sharply: Recovery Will Be Duplicated in East, Executives Predict." *New York Times,* March 17, 1947. 43.

Allan, J. McGrigor. "On the Real Differences in the Minds of Men and Women." *Journal of the Anthropological Society of London* 749, no. 16 (1869): cxcv–ccxix.

Allen, Danielle S. *Talking to Strangers: Anxieties of Citizenship since Brown v. Board of Education.* Chicago: University of Chicago Press, 2004.

Amen, Daniel G. *The Brain in Love: 12 Lessons to Enhance Your Love Life.* New York: Three Rivers Press, 2009.

Ames, Melissa. "Exploding Rhetorics of 9/11: An Approach for Studying the Role That Affect and Emotion Play in Constructing Historical Events." *Pedagogy* 17, no. 2 (2017): 177–202.

Amodio, David M., John T. Jost, Sarah L. Master, and Cindy M. Yee. "Neurocognitive Correlates of Liberalism and Conservatism." *Nature Neuroscience* 10, no. 10 (2007): 1246–47.

Aristotle. *The Nicomachean Ethics.* Translated by Harris Rackham. Loeb Classical Library 73. Cambridge, MA: Harvard University Press, 2009.

———. *Rhetoric.* Translated by W. Rhys Roberts. New York: Modern Library, 1954.

Baake, Ken. *Metaphor and Knowledge: The Challenges of Writing Science.* Albany: State University of New York Press, 2003.

Bain, Alexander. *English Composition and Rhetoric.* New York: D. Appleton and Company, 1890.

———. *The Senses and the Intellect.* London: John W. Parker and Son, 1855.

———. *The Senses and the Intellect.* 3rd ed. New York: D. Appleton and Company, 1868.

Baker, D. A., Jillian M. Ware, N. J. Schweitzer, and Evan F. Risko. "Making Sense of Research on the Neuroimage Bias." *Public Understanding of Science* 26, no. 2 (2017): 251–58.

Barad, Karen Michelle. *Meeting the Universe Halfway: Quantum Physics and the Entanglement of Matter and Meaning.* Durham, NC: Duke University Press, 2007.

Barclay, Michael W. "The Metaphoric Foundation of Literal Language: Towards a Theory of the Reification and Meaning of Psychological Constructs." *Theory & Psychology* 7, no. 3 (1997): 355–72.

Barr, Donald. "A Decade of Miracles, Once Over Lightly." *New York Times,* June 15, 1947.

Barrett, Lisa Feldman. "Navigating the Science of Emotion." *Emotion Measurement,* January 1, 2016, 31–63. doi:10.1016/B978-0-08-100508-8.00002-3.

———. "The Theory of Constructed Emotion: An Active Inference Account of Interoception and Categorization." *Social Cognitive and Affective Neuroscience* 12, no. 1 (2017): 1–23.

Barrett, Lisa Feldman, and Ajay B. Satpute. "Historical Pitfalls and New Directions in the Neuroscience of Emotion." *Neuroscience Letters,* 2017. doi:10.1016/j.neulet.2017.07.045.

Battle, Craig. "Remembering Tragically Hip's Contributions to Hockey Night in Canada." *Sportsnet,* October 18, 2017. http://www.sportsnet.ca/hockey/nhl/remembering-tragically-hips-contributions-hockey-night-canada/.

Baudrillard, Jean. "Simulacra and Simulations." In *Selected Writings,* edited by Mark Poster, 169–87. Stanford, CA: Stanford University Press, 2001.

Bawarshi, Anis. "The Genre Function." *College English* 62, no. 3 (2000): 335–60.

Beam, Elizabeth, L Gregory Appelbaum, Jordynn Jack, James Moody, and Scott A Huettel. "Mapping the Semantic Structure of Cognitive Neuroscience," *Journal of Cognitive Neuroscience* 26, no. 9 (2011): 1949–65.

Beer, David. "The Future of Affect Theory: An Interview with Margaret Wetherell." *Theory, Culture & Society,* October 15, 2014. https://www.theoryculturesociety.org/the-future-of-affect-theory-an-interview-with-margaret-wetherall/.

Bergson, Henri. *Time and Free Will: An Essay on the Immediate Data of Consciousness.* Translated by F. L. Pogson. Mineola, NY: Dover, 2001.

Bickford, Susan. "Constructing Inequality: City Spaces and the Architecture of Citizenship." *Political Theory* 28, no. 3 (2000): 355–76.

Bitzer, Lloyd F. "Functional Communication: A Situational Perspective." In *Rhetoric in Transition: Studies in the Nature and Uses of Rhetoric,* edited by Eugene Edmond White, 21–38. University Park: Pennsylvania State University Press, 1980.

———. "The Rhetorical Situation." *Philosophy and Rhetoric* 1, no. 1 (1968): 1–14.

Bizzell, Patricia, and Bruce Herzberg, eds. *The Rhetorical Tradition: Readings from Classical Times to the Present.* 2nd ed. Boston: Bedford/St. Martin's, 2001.

Blakemore, Erin. "Unraveling the Mysteries of the Addicted Brain." *The Washington Post,* September 2, 2017.

Bonnet, Charles. *La palingénésie philosophique.* Geneva: Claude Philiburt et Barthelemi Chirol, 1769.

Booth, Wayne C. *The Rhetoric of Fiction.* Chicago: University of Chicago Press, 1983.

Borg, Charmaine, Peter J. de Jong, and Janniko R. Georgiadis. "Subcortical BOLD Responses during Visual Sexual Stimulation Vary as a Function of Implicit Porn Associations in Women." *Social Cognitive and Affective Neuroscience* 9, no. 2 (2014): 158–66.

Broca, Paul. "Remarks on the Seat of Spoken Language, Followed by a Case of Aphasia (1861)." *Neuropsychology Review* 21, no. 3 (2011): 227–29.

———. "Remarks on the Seat of the Faculty of Articulated Language." Translated by Christopher D. Green. *Classics in the History of Psychology.* Published 2000. http://psychclassics.yorku.ca/Broca/aphemie-e.htm.

Brown, JoAnne. *The Definition of a Profession: The Authority of Metaphor in the History of Intelligence Testing, 1890–1930.* Princeton, NJ: Princeton University Press, 1992.

Brueggemann, Brenda. "Whole Brains, Half-Brains, and Writing." *Rhetoric Review* 8, no. 1 (1989): 127–36.

Burke, Kenneth. *Counter-Statement*. Berkeley: University of California Press, 1958.

———. "Four Master Tropes." *The Kenyon Review* 3, no. 4 (1941): 421–38.

———. *Language as Symbolic Action*. Berkeley: University of California Press, 1966.

———. *Permanence and Change*. 3rd ed. Berkeley: University of California Press, 1954.

———. *A Rhetoric of Motives*. Berkeley: University of California Press, 1969.

Burnett, Dean. "How 'Provocative Clothes' Affect the Brain—And Why It's No Excuse for Assault." *The Guardian,* January 25, 2018. https://www.theguardian.com/science/brain-flapping/2018/jan/25/how-provocative-clothes-affect-the-brain-and-why-its-no-excuse-for-assault.

Cabanis, Jean Pierre George. *Rapports du physique et du moral de l'homme*. 3rd ed. 2 vols. Paris, 1815.

Campbell, George. *The Philosophy of Rhetoric*. 7th ed. London: William Haynes and Son, 1823.

Carter, Michael. "Stasis and Kairos: Principles of Social Construction in Classical Rhetoric." *Rhetoric Review* 7, no. 1 (1988): 97–112.

Caufield, Timothy, Christen Rachul, and Amy Zarzeczny. "'Neurohype' and the Name Game: Who's to Blame?" *AJOB Neuroscience* 1, no. 2 (2010): 13–40.

Charland, Maurice. "Constitutive Rhetoric: The Case of the *Peuple Québécois*." *Quarterly Journal of Speech* 73, no. 2 (1987): 133–50.

Chaturvedi, Richa. "A Closer Look at the Gender Gap in Presidential Voting." *Pew Research Center,* July 28, 2016. http://www.pewresearch.org/fact-tank/2016/07/28/a-closer-look-at-the-gender-gap-in-presidential-voting/.

Chivers, Meredith L., Michael C. Seto, and Ray Blanchard. "Gender and Sexual Orientation Differences in Sexual Response to Sexual Activities Versus Gender of Actors in Sexual Films." *Journal of Personality and Social Psychology* 93, no. 6 (2007), 1108–21.

Ciclitira, Karen. "Pornography, Women and Feminism: Between Pleasure and Politics." *Sexualities* 7, no. 3 (August 17, 2004): 281–301.

Cikara, Mina, Jennifer L. Eberhardt, and Susan T. Fiske. "From Agents to Objects: Sexist Attitudes and Neural Responses to Sexualized Targets." *Journal of Cognitive Neuroscience* 23, no. 3 (2011): 540–51.

Cisneros, Josue David. "Looking 'Illegal': Affect, Rhetoric, and Performativity in Arizona's Senate Bill 1070." In *Border Rhetorics: Citizenship and Identity on the US-Mexico Frontier,* edited by D. Robert DeChaine, 133–50. Tuscaloosa: University of Alabama Press, 2012.

Clark, Andy. "Dreaming the Whole Cat: Generative Models, Predictive Processing, and the Enactivist Conception of Perceptual Experience." *Mind* 121, no. 483 (2012): 753–71.

———. *Supersizing the Mind: Embodiment, Action, and Cognitive Extension*. Oxford: Oxford University Press, 2011.

Clarke, Edwin, and L. S. Jacyna. *Nineteenth-Century Origins of Neuroscientific Concepts*. Berkeley: University of California Press, 1987.

Cohn, Simon. "Petty Cash and the Neuroscientific Mapping of Pleasure." *BioSocieties* 3, no. 2 (2008): 151–63.

Condit, Celeste Michelle. *Decoding Abortion Rhetoric: Communicating Social Change*. Urbana: University of Illinois Press, 1990.

———. "How Bad Science Stays That Way: Brain Sex, Demarcation, and the Status of Truth in the Rhetoric of Science." *Rhetoric Society Quarterly* 26, no. 4 (2014): 83–109.

Cooper, Anderson. "60 Minutes." CBS, 2011. https://www.youtube.com/watch?v=MLtd_G90Zvo.

Corbett, Edward P. J. "The Theory and Practice of Imitation in Classical Rhetoric." *College Composition and Communication* 22, no. 3 (1971): 243–50.

Crary, Jonathan. *Techniques of the Observer: On Vision and Modernity in the Nineteenth Century.* Cambridge, MA: MIT Press, 1992.

Crenshaw, Kimberlé. "Mapping the Margins: Intersectionality, Identity Politics, and Violence Against Women of Color." *Stanford Law Review* 43, no. 6 (1991): 1241–99.

Cronin, Melissa. "Rappers' Brains: Relaxed 'Executive Function' May Enable Freestyle Raps." *Huffington Post,* December 6, 2017. https://www.huffingtonpost.com/2012/11/16/rappers-brains-executive-function-freestyle_n_2144874.html

Crowley, Sharon. *Toward a Civil Discourse: Rhetoric and Fundamentalism.* Pittsburgh, PA: University of Pittsburgh Press, 2006.

Damasio, Antonio R. "The Somatic Marker Hypothesis and the Possible Functions of the Prefrontal Cortex." *Philosophical Transactions of the Royal Society B: Biological Sciences* 351, no. 1346 (1996): 1413–20.

Descartes, René. *L'homme.* Paris, 1664.

Dhamoon, Rita Kaur. "Considerations on Mainstreaming Intersectionality." *Political Research Quarterly* 64, no. 1 (2011): 230–43.

Diamond, Jeremy. "Donald Trump: Ban All Muslim Travel to U.S." *CNN,* December 8, 2015. http://www.cnn.com/2015/12/07/politics/donald-trump-muslim-ban-immigration/.

Dickinson, Emily. "The Lost Thought," in *The Poems of Emily Dickinson,* edited by Thomas H. Johnson, Cambridge, Mass.: The Belknap Press of Harvard University Press, Copyright © 1951, 1955 by the President and Fellows of Harvard College. Copyright © renewed 1979, 1983 by the President and Fellows of Harvard College. Copyright © 1914, 1918, 1919, 1924, 1929, 1930, 1932, 1935, 1937, 1942, by Martha Dickinson Bianchi. Copyright © 1952, 1957, 1958, 1963, 1965, by Mary L. Hampson.

Dodd, Michael D., Amanda Balzer, Carly M. Jacobs, Michael W. Gruszczynski, Kevin B. Smith, and John R. Hibbing. "The Political Left Rolls with the Good and the Political Right Confronts the Bad: Connecting Physiology and Cognition to Preferences." *Philosophical Transactions: Biological Sciences* 367, no. 1589 (2012): 640–49.

Donald J. Trump for President. "Immigration Reform." 2016. https://www.donaldjtrump.com/positions/immigration-reform. Retrieved from archive.org.

Driskill, Qwo-Li. "Decolonial Skillshares: Indigenous Rhetorics as Radical Practice." In *Survivance, Sovereignty, and Story: Teaching American Indian Rhetorics,* edited by Lisa King, Rose Gubele, and Joyce Rain Anderson, 57–78. Logan: Utah State University Press, 2015.

Dussauge, Isabelle. "The Experimental Neuro-Framing of Sexuality." *Graduate Journal of Social Science* 10, no. 1 (2013): 124–51.

Edbauer, Jenny. "Unframing Models of Public Distribution: From Rhetorical Situation to Rhetorical Ecologies." *Rhetoric Society Quarterly* 35, no. 4 (2005): 5–24.

"The Electronic and the Human Brain." *Lancet* 248, no. 6431 (1946): 795–96.

"Electronic Brain." *British Medical Journal* 2, no. 4482 (1946): 824.

"Electronic Brain Does Research." *New York Times,* July 3, 1949. E7.

"Electronic Control Speeds Production of Ship Propellers." *Marine Engineering and Shipping Review* 50, no. 7 (1952): 137.

Eminem. Interview by Anderson Cooper. *60 Minutes*. CBS. October 10, 2010. https://www.youtube.com/watch?v=MLtd_G9oZvo.

Engels, Jeremy. *The Politics of Resentment: A Genealogy*. University Park: Pennsylvania State University Press, 2015. Kindle.

Entman, Robert M. "Framing: Toward Clarification of a Fractured Paradigm." *Journal of Communication* 43, no. 4 (1993): 51–58.

Epstein, Robert. "Your Brain Does Not Process Information and It Is Not a Computer." *Aeon*, May 18, 2016. https://aeon.co/essays/your-brain-does-not-process-information-and-it-is-not-a-computer.

Fahnestock, Jeanne. "Accommodating Science: The Rhetorical Life of Scientific Facts" *Written Communication* 15, no. 3 (1998): 330–50.

———. "Rhetoric in the Age of Cognitive Science." In *The Viability of the Rhetorical Tradition*, edited by Richard Graff, Arthur Walzer, and Janet Atwill, 159–79. Albany: State University of New York Press, 2005.

———. *Rhetorical Figures in Science*. New York: Oxford University Press, 1999.

———. *Rhetorical Style: The Uses of Language in Persuasion*. New York: Oxford University Press, 2011.

Fahnestock, Jeanne, and Marie Secor. "The Stases in Scientific and Literary Argument." *Written Communication* 5, no. 4 (1988): 427–43.

Falk, Emily B., Elliot T. Berkman, Traci Mann, Brittany Harrison, and Matthew D. Lieberman. "Predicting Persuasion-Induced Behavior Change from the Brain." *Journal of Neuroscience* 30, no. 25 (2010): 8421–24.

Farah, Martha J., and Cayce J. Hook. "The Seductive Allure of 'Seductive Allure.'" *Perspectives on Psychological Science* 8, no. 1 (2015): 2013–15.

Farrell, Thomas B. "Practicing the Arts of Rhetoric: Tradition and Invention." *Philosophy and Rhetoric* 24, no. 3 (2015): 183–212.

Fausto-Sterling, Anne. *Sexing the Body: Gender Politics and the Construction of Sexuality*. New York: Basic Books, 2000.

Fausto-Sterling, Anne, Cynthia Garcia Coll, and Meaghan Lamarre. "Sexing the Baby, Part 2: Applying Dynamic Systems Theory to the Emergences of Sex-Related Differences in Infants and Toddlers." *Social Science and Medicine* 74, no. 11 (2012): 1693–702.

Fernandez-Duque, Diego, Jessica Evans, Colton Christian, and Sara D. Hodges. "Superfluous Neuroscience Information Makes Explanations of Psychological Phenomena More Appealing." *Journal of Cognitive Neuroscience* 27, no. 5 (2015): 926–44.

Ferrier, David. *The Functions of the Brain*. 2nd ed. New York: G.P. Putnam's Sons, 1886.

Fine, Cordelia. *Delusions of Gender: How Our Minds, Society, and Neurosexism Create Difference*. New York: W.W. Norton, 2010.

Fink, Andreas, Barbara Graif, and Aljoscha C. Neubauer. "Brain Correlates Underlying Creative Thinking: EEG Alpha Activity in Professional vs. Novice Dancers." *NeuroImage* 46, no. 3 (2009): 854–62.

Finnegan, Cara A. "The Naturalistic Enthymeme and Visual Argument: Photographic Representation in the 'Skull Controversy.'" *Argumentation and Advocacy* 37, no. 3 (2001): 133–49.

Fleming, David. "Rhetoric as a Course of Study." *College English* 61, no. 2 (1998): 169–91.

Flourens, P. *Recherches expérimentales sur les propriétés et les fonctions du système nerveux dans les animaux vertébrés.* Paris, 1842.

Flower, Linda, and John R Hayes. "A Cognitive Process Theory of Writing." *College Composition and Communication* 32, no. 4 (1981): 365. doi:10.2307/356600.

"Full Text: Donald Trump 2016 RNC Draft Speech Transcript." *Politico,* July 21, 2016. https://www.politico.com/story/2016/07/full-transcript-donald-trump-nomination-acceptance-speech-at-rnc-225974.

Funk, Cary, and Lee Rainie. "Americans, Politics and Science Issues." *Pew Research Center,* July 1, 2015. http://assets.pewresearch.org/wp-content/uploads/sites/14/2015/07/2015-07-01_science-and-politics_FINAL-1.pdf.

Gall, Franz Joseph. *On the Function of the Brain and of Each of Its Parts.* Translated by Winslow Lewis. 6 vols. Boston, 1835.

———. *Sur l'origine des qualités morales et des facultés intellectuelles de l'homme.* 6 vols. Paris, 1822–25.

Gall, Franz Joseph, and Johann Caspar Spurzheim. *Anatomie et physiologie du système nerveux.* 4 vols. Paris, 1810–19.

Gallo, Carmine. "Neuroscientists Say This Tool Is the Secret to Persuasion." *Inc.com,* October 25, 2017. https://www.inc.com/carmine-gallo/your-single-best-tool-to-instantly-click-with-another-person-according-to-cognitive-scientists.html (accessed June 18, 2018).

Gallup. "Abortion." *Gallup News,* n.d. http://www.gallup.com/poll/1576/abortion.aspx (accessed April 19, 2017).

Garner, Ruth, Mark G. Gillingham, and C . Stephen White. "Effects of 'Seductive Details' on Macroprocessing and Microprocessing in Adults and Children." *Cognition and Instruction* 6, no. 1 (1989): 41–57.

Gibbons, Michelle. "A Neurorhetoric of Incongruity." *POROI: An Interdisciplinary Journal of Rhetorical Analysis and Invention* 13, no. 2 (2018).

Glick, Peter, and Susan T. Fiske. "The Ambivalent Sexism Inventory: Differentiating Hostile and Benevolent Sexism." *Journal of Personality and Social Psychology* 70, no. 3 (1996): 491–512.

Goldman, Alvin, and Frederique de Vignemont. "Is Social Cognition Embodied?" *Trends in Cognitive Sciences* 13, no. 4 (2009): 154–59.

Gorgias. *Encomium of Helen.* Translated by George A. Kennedy. In Bizzell and Herzberg, *The Rhetorical Tradition,* 44–46.

Gower, John. *Confessio Amantis.* Edited by Reinhold Pauli. London, 1857.

Green Party. "Preamble." *Green Party US,* 2016. http://www.gp.org/preamble_2016.

———. "Ten Key Values." *Green Party US,* 2016. http://www.gp.org/ten_key_values_2016.

Gross, Daniel M. *The Secret History of Emotion: From Aristotle's Rhetoric to Modern Brain Science.* Chicago: University of Chicago Press, 2006.

Gross, Daniel M., and Stephanie D. Preston. "Emotion Science and the Heart of a Two-Cultures Problem." In *Science and Emotions after 1945: A Transatlantic Perspective,* edited by Frank Biess and Daniel M. Gross, 96–117. Chicago: University of Chicago Press, 2014.

Gruber, David. "The Extent of Engagement, the Means of Invention: Measuring Debate about Mirror Neurons in the Humanities and Social Sciences." *Journal of Science Communication* 15, no. 2 (2016): 1–17.

———. "The Macro-Micro." *NeuroHuman.com,* n.d. http://www.neurohuman.com/project/macro/ (accessed July 17, 2017).

———. "The Neuroscience of Rhetoric : Identification, Mirror Neurons, and Making the Many Appear." In *Neurorhetorics,* edited by Jordynn Jack, 37–52. New York: Routledge, 2013.

———. "Reinventing the Brain, Revising Neurorhetorics: Phenomenological Networks Contesting Neurobiological Interpretations." *Rhetoric Review* 35, no. 3 (2016): 239–53.

———. "Three Forms of Neuro-Realism: Explaining the Persistence of the 'Uncritically Real' in Popular Neuroscience News." *Written Communication* 34, no. 2 (2017): 189–223.

Guilford, J. P. "Creativity." *American Psychologist* 5, no. 9 (1950): 444–54.

Hamann, Stephan, Rebecca A. Herman, Carla L. Nolan, and Kim Wallen. "Men and Women Differ in Amygdala Response to Visual Sexual Stimuli." *Nature Neuroscience* 7, no. 4 (2004): 411–16.

Hamilton, Cynthia S. "'Am I Not a Man and a Brother?' Phrenology and Anti-Slavery." *Slavery and Abolition* 29, no. 2 (2008): 173–87.

Hancock, Ange-Marie. *The Politics of Disgust: The Public Identity of the Welfare Queen.* New York: New York University Press, 2004.

Haraway, Donna. "Situated Knowledges: The Science Question in Feminism and the Privilege of Partial Perspective." *Feminist Studies* 14, no. 3 (1988): 575–99.

Harris, Randy Allen. "Introduction." In *Rhetoric and Incommensurability,* 3–149. West Lafayette, IN: Parlor Press, 2005.

Hart, Peter D. "In 2016 Voter's Anxiety and Fear, a Message to the Establishment." *Wall Street Journal,* January 26, 2016. https://blogs.wsj.com/washwire/2016/01/26/in-2016-voters-anxiety-and-fear-a-message-to-the-establishment/.

Hawhee, Debra. *Bodily Arts: Rhetoric and Athletics in Ancient Greece.* Austin: University of Texas Press, 2004.

———. "Burke on Drugs." *Rhetoric Society Quarterly* 34, no. 1 (2004): 5–28.

Hayles, N. Katherine. *How We Became Posthuman: Virtual Bodies in Cybernetics, Literature, and Informatics.* Chicago: University of Chicago Press, 1999.

Hetherington, Marc J., and Jonathan Weiler. *Authoritarianism and Polarization in American Politics.* Cambridge: Cambridge University Press, 2009.

"Hillary Clinton's DNC Speech." *CNN.com,* July 29, 2016. https://www.cnn.com/2016/07/28/politics/hillary-clinton-speech-prepared-remarks-transcript/index.html.

Hippocrates. "The Sacred Disease." In *Hippocrates,* vol. 2. Translated by W. H. S. Jones. Loeb Classical Library 148. Cambridge, MA: Harvard University Press, 1923.

Hohmann, Marti. "The Politics of the G-Spot: Penetrative Sex in Feminist Pornography, 1981–1991." *Harvard Gay & Lesbian Review* 3, no. 3 (1996): 23–25.

Holmes, David G. "Affirmative Reaction: Kennedy, Nixon, King, and the Evolution of Color-Blind Rhetoric." *Rhetoric Review* 26, no. 1 (2007): 25–41.

Hunter, George. *How to Defend Yourself, Your Family, and Your Home: A Complete Guide to Self-Protection.* New York: David McKay, 1968.

Ingold, Tim. *Being Alive: Essays on Movement, Knowledge and Description.* London: Routledge, 2011.

———. *Making: Anthropology, Archaeology, Art, and Architecture.* New York: Routledge, 2013.

———. "The Textility of Making." *Cambridge Journal of Economics* 34, no. 1 (2010): 91–102.

Irlande, John. *The Meroure of Wyssdome: Composed for the Use of James IV., King of Scots, A.D. 1490. Scottish Text Society, 4th Ser., 19.* Vol. III. Edinburgh: Printed for the Society by W. Blackwood, 1990.

Isocrates. "Antidosis." Translated by George Norlin. In Bizzell and Herzberg, *The Rhetorical Tradition*, 75–79.

Jack, Jordynn. "'The Piety of Degradation': Kenneth Burke, the Bureau of Social Hygiene, and *Permanence and Change*." *Quarterly Journal of Speech* 90, no. 4 (2004): 446–68.

Jack, Jordynn, and L. Gregory Appelbaum. "'This Is Your Brain on Rhetoric': Research Directions for Neurorhetorics." *Rhetoric Society Quarterly* 40, no. 5 (2010): 411–37.

Jack, Jordynn, L. Gregory Appelbaum, Elizabeth Beam, James Moody, and Scott A Huettel. "Mapping Rhetorical Topologies in Cognitive Neuroscience." In *Topologies as Techniques for a Post-Critical Rhetoric,* edited by Lynda Walsh and Casey Boyle, 125–50. New York: Palgrave Macmillan, 2017.

Jaffe, Eric. "My Brain Is a Walnut: Inside an fMRI Machine." *Slate,* January 10, 2006. https://www.slate.com/articles/health_and_science/science/2006/01/my_brain_is_a_walnut.html.

Jamieson, Jeremy P., Wendy Berry Mendes, Erin Blackstock, and Toni Schmader. "Turning the Knots in Your Stomach into Bows: Reappraising Arousal Improves Performance on the GRE." *Journal of Experimental Social Psychology* 46, no. 1 (2009): 208–12.

Jay-Z. *Decoded.* New York: Spiegel & Grau, 2011.

Jensen, Robin E. "An Ecological Turn in Rhetoric of Health Scholarship: Attending to the Historical Flow and Percolation of Ideas, Assumptions, and Arguments." *Communication Quarterly* 63, no. 5 (2015): 522–26.

Joel, Daphna. "Male or Female? Brains Are Intersex." *Frontiers in Integrative Neuroscience* 5, (2011). doi:10.3389/fnint.2011.00057.

Joel, Daphna, Zohar Berman, Ido Tavor, Nadav Wexler, Olga Gaber, Yaniv Stein, Nisan Shefi, Jared Pool, Sebastian Urchs, Daniel S. Margulies, Franziskus Liem, Jürgen Hänggi, Lutz Jäncke, and Yaniv Assaf. "Sex beyond the Genitalia: The Human Brain Mosaic." *Proceedings of the National Academy of Sciences* 112, no. 50 (2015): 15468–73.

Johnson, Jenell M., and Melissa M. Littlefield. "Lost and Found in Translation: Popular Neuroscience in the Emerging Neurodisciplines." In *Sociological Reflections on the Neurosciences,* edited by Martyn Pickersgill and Ira Van Keulen, 279–97. Bingley: Emerald, 2011.

Johnson, Robert R. *User-Centered Technology: A Rhetorical Theory for Computers and Other Mundane Artifacts.* Albany: State University of New York Press, 1998.

Jordan, Heather, Mazyar Fallah, and Gene R. Stoner. "Adaptation of Gender Derived from Biological Motion." *Nature Neuroscience* 9, no. 6 (2006): 738–39.

Jordan-Young, Rebecca M. *Brain Storm: The Flaws in the Science of Sex Differences.* Cambridge: Harvard University Press, 2010.

Kaite, Berkeley. *Pornography and Difference.* Bloomington: Indiana University Press, 1995.

Kanai, Ryota, Tom Feilden, Colin Firth, and Geraint Rees. "Political Orientations Are Correlated with Brain Structure in Young Adults." *Current Biology* 21, no. 8 (2011): 677–80.

Katsurakawa, Motomu, and Katsuyuki Sakai. "Unraveling Brain Network Coding with a Connectivity-Based Classifier." *Trends in Cognitive Sciences* 16, no. 10 (2012): 492–94.

Keehner, Madeleine, Lisa Mayberry, and Martin H. Fischer. "Different Clues from Different Views: The Role of Image Format in Public Perceptions of Neuroimaging Results." *Psychonomic Bulletin & Review* 18, no. 2 (2011): 422–28.

Keller, Katherine, and Vinod Menon. "Gender Differences in the Functional and Structural Neuroanatomy of Mathematical Cognition." *NeuroImage* 47, no. 1 (2009): 342–52.

Klucharev, Vasily, Ale Smidts, and Guillén Fernández. "Brain Mechanisms of Persuasion: How 'Expert Power' Modulates Memory and Attitudes." *Social Cognitive and Affective Neuroscience* 3, no. 4 (2008): 353–66.

Knight, Robert T. "Neural Networks Debunk Phrenology." *Science* 316, no. 5831 (2007): 1578–79.

Koedt, Anne. "The Myth of the Vaginal Orgasm." In *Radical Feminism,* edited by Anne Koedt, Ellen Levine, and Anita Rapone, 198–207. New York: Quadrangle Books, 1973.

Kühn, Simone, Enrique Strelow, and Jürgen Gallinat. "Multiple 'Buy Buttons' in the Brain: Forecasting Chocolate Sales at Point-of-Sale Based on Functional Brain Activation Using FMRI." *NeuroImage* 136 (August 1, 2016): 122–28.

Lange, Jeva. "American *Pravda*: How Donald Trump Could Sovietize the Media." *The Week,* November 28, 2016. https://theweek.com/articles/662731/american-pravda-how-donald-trump-could-sovietize-media.

Latour, Bruno. *Science in Action: How to Follow Scientists and Engineers through Society.* Cambridge, MA: Harvard University Press, 1987.

Latour, Bruno, and Steve Woolgar. *Laboratory Life: The Construction of Scientific Facts.* Princeton, NJ: Princeton University Press, 1986.

Lawson, George. *Theo-Politica: Or, a Body of Divinity.* London: The second edition, corrected. London, 1705. Eighteenth Century Collections Online. Gale. 11 Jan. 2019.

Leary, David. *Metaphors in the History of Psychology.* New York: Cambridge University Press, 1990.

Lee, Michael J. *Creating Conservatism: Postwar Words That Made an American Movement.* East Lansing: Michigan State University Press, 2014.

LeFevre, Karen Burke. *Invention as a Social Act.* Carbondale: Southern Illinois University Press, 1986.

Legrenzi, Paolo, and Carlo Umiltà. *Neuromania: On the Limits of Brain Science.* Oxford: Oxford University Press, 2011.

Leknes, Siri, and Irene Tracey. "A Common Neurobiology for Pain and Pleasure." *Nature Reviews Neuroscience* 9, no. 4 (2008): 314–20.

Leys, Ruth. "The Turn to Affect: A Critique." *Critical Inquiry* 37, no. 3 (2011): 434–72.

Libertarian Party. "2018 Platform." *Libertarian Party,* July 2018. https://www.lp.org/platform/.

Lil Yachty. "Llama Llama Red Pajama over Ms. Jackson Beat." *The Cruz Show,* Power 106 Los Angeles, May 24, 2017. https://www.youtube.com/watch?v=e6OLsNKBfDE.

Lindquist, Kristen A., Maria Gendron, and Ajay B. Satpute. "Language and Emotion: Putting Words into Feelings and Feelings into Words." In *Handbook of Emotions,* 4th ed., edited by Lisa Feldman Barrett, Michael Lewis, and Jeannette M. Haviland-Jones, 579–94. New York: Guilford Press, 2016.

Liu, Siyuan, Ho Ming Chow, Yisheng Xu, Michael G. Erkkinen, Katherine E. Swett, Michael W. Eagle, Daniel A. Rizik-Baer, and Allen R. Braun. "Neural Correlates of Lyrical Improvisation: An fMRI Study of Freestyle Rap." *Scientific Reports* 2, no. 834 (2012): 1–8.

Lobb, Andrea. "The Agony and the Empathy: The Ambivalence of Empathy in Feminist Psychology." *Feminism & Psychology* 23, no. 4 (2013): 426–41.

Locke, John. *An Essay Concerning Human Understanding.* 3 vols. Edinburgh, 1798.

Ludacris. "Llama Llama Red Pajama Freestyle." *The Cruz Show,* Power 106 Los Angeles, April 13, 2017. https://www.youtube.com/watch?v=PFtHeo7oMSU.

Lupyan, Gary. "Linguistically Modulated Perception and Cognition: The Label-Feedback Hypothesis." *Frontiers in Psychology* 3, no. 54 (2012): 1–13.

Majlis Ash-Shura. "4th Term 3rd Year, 1428–1429 A. H." *Shura Council, Kingdom of Saudi Arabia,* 2007. https://shura.gov.sa/wps/wcm/connect/ShuraEn/internet/Royal+Speeches/4th+Term+3rd+Year,+1428+-+1429+A.H./

Margulies, Daniel S. "The Salmon of Doubt." In *Critical Neuroscience: A Handbook of the Social and Cultural Contexts of Neuroscience,* edited by Suparna Choudhury and Jan Slaby, 273–85. Oxford: Wiley-Blackwell, 2011.

———. "Unraveling the Complex Tapestry of Association Networks." *Neuron* 95, no. 109 (2017): 239–41.

Massumi, Brian. "The Autonomy of Affect." *Cultural Critique,* no. 31 (1995): 83–109.

May, Caroline. "Poll: Majority of Voters Fear 'Homegrown Jihadists,' Most Oppose Accepting Syrian Refugees." *Breitbart,* December 3, 2015. http://www.breitbart.com/big-government/2015/12/03/poll-majority-voters-fear-homegrown-jihadists-oppose-accepting-syrian-refugees/.

Mays, Chris, and Julie Jung. "Priming Terministic Inquiry: Toward a Methodology of Neurorhetoric." *Rhetoric Review* 31, no. 1 (2012): 41–59.

McCabe, David P., and Alan D. Castel. "Seeing Is Believing: The Effect of Brain Images on Judgments of Scientific Reasoning." *Cognition* 107, no. 1 (2008): 343–52.

McDougall, W. *Physiological Psychology.* London: J. M. Dent, 1905.

Merenda, Peter F. "Measurements in the Future: Beyond the 20th Century." *Psychological Reports* 92, no. 1 (2003): 209–17.

Miller, Carolyn R. "The Aristotelian Topos: Hunting for Novelty." In *Rereading Aristotle's Rhetoric,* edited by Alan G. Gross and Arthur E. Walzer, 130–46. Carbondale: Southern Illinois University Press, 2000.

Mitchell, Jason P., Daniel L. Ames, Adrianna C. Jenkins, and Mahzarin R. Banaji. "Neural Correlates of Stereotype Application." *Journal of Cognitive Neuroscience* 21, no. 3 (2009): 594–604.

Moore, Jessie L, Paula Rosinski, Tim Peeples, Stacey Pigg, Martine Courant Rife, Beth Brunk-Chavez, Dundee Lackey, et al. "Revisualizing Composition: How First-Year Writers Use Composing Technologies." *Computers and Composition* 39, no. 39 (2016): 1–13.

Mouffe, Chantal. *On the Political.* London: Routledge, 2005.

"The Mt. Oread Manifesto on Rhetorical Education 2013." *Rhetoric Society Quarterly* 44, no. 1 (2014): 1–5.

Muckelbauer, John. *The Future of Invention: Rhetoric, Postmodernism, and the Problem of Change.* Albany: State University of New York Press, 2009.

Mudry, Jessica J. *Measured Meals: Nutrition in America.* Albany: SUNY Press, 2009.

Muller, Daphne. "Your Brain Isn't a Computer—It's a Quantum Field." *Big Think,* September 18, 2017. https://bigthink.com/ideafeed/does-the-mind-play-dice-with-reason.

Murphy, James Jerome. *A Short History of Writing Instruction: From Ancient Greece to Contemporary America.* New York: Routledge, 2012.

"New Proposals on Abortion." *Life,* March 1967.

Newcomb, L. H. "Toward a Sustainable Source Study." In *Rethinking Shakespeare Source Study: Audiences, Authors, and Digital Technologies,* edited by Dennis Austin Britton and Melissa Walter. New York: Routledge, 2018. https://doi.org/10.4324/9781315649061-9

Newman, M. H. A. "A Note on Electric Automatic Computing Machines." *British Medical Journal* 1, no. 4616 (1949): 1133.

Noë, Alva. *Out of Our Heads: Why You Are Not Your Brain, and Other Lessons from the Biology of Consciousness*. New York: Hill and Wang, 2009.

O'Brien, Kerry, Walter Forrest, Dermot Lynott, and Michael Daly. "Racism, Gun Ownership and Gun Control: Biased Attitudes in US Whites May Influence Policy Decisions." *PLoS ONE* 8, no. 10 (2013): e77552.

Occupy Wall Street. "Declaration." http://www.occupywallstreet.net/learn (accessed July 5, 2017).

Ofengeim, Dimitry, Nikolaos Giagtzoglou, Dann Huh, Chengyu Zou, and Junying Yuan. "Single-Cell RNA Sequencing: Unraveling the Brain One Cell at a Time." *Trends in Molecular Medicine* 23, no. 6 (2017): 563–76.

Olds, James. "Pleasure Centers in the Brain." *Scientific American* 195, no. 4 (1956): 105–17.

———. "Self-Stimulation of the Brain." *Science* 127, no. 3294 (1958): 315–24.

Olds, James, and Peter Milner. "Positive Reinforcement Produced by Electrical Stimulation of Septal Area and Other Regions of Rat Brain." *Journal of Comparative and Physiological Psychology* 47, no. 6 (1954): 419–27.

Onians, Richard Broxton. *The Origins of European Thought: About the Body, the Mind, the Soul, the World, and Fate*. Cambridge: Cambridge University Press, 1951.

Ott, Brian L. "The Visceral Politics of *V for Vendetta*: On Political Affect in Cinema." *Critical Studies in Media Communication* 27, no. 1 (2010): 39–54.

Panda, R., et al. "Unraveling Brain Functional Connectivity of Encoding and Retrieval in the Context of Education." *Brain and Cognition* 86, no. 1 (2014): 75–81.

Papoulias, Constantina, and Felicity Callard. "Biology's Gift: Interrogating the Turn to Affect." *Body & Society* 16, no. 1 (2010): 29–56.

Park, David. "Picturing the War: Visual Genres in Civil War News." *Communication Review* 3, no. 4 (1999): 287–321.

Petty, Richard E., and John T. Cacioppo. "The Elaboration Likelihood Model of Persuasion." *Advances in Experimental Social Psychology* 19 (1986): 123–205.

Pew Research Center. "83% Say Measles Vaccine Is Safe for Healthy Children." *Pew Research Center*, February 9, 2015. http://www.people-press.org/2015/02/09/83-percent-say-measles-vaccine-is-safe-for-healthy-children/.

———. "A Deep Dive Into Party Affiliation." *Pew Research Center*, April 7, 2015. http://www.people-press.org/2015/04/07/a-deep-dive-into-party-affiliation/.

———. "The Parties on the Eve of the 2016 Election: Two Coalitions, Moving Further Apart." *Pew Research Center*, September 13, 2016. http://www.people-press.org/2016/09/13/the-parties-on-the-eve-of-the-2016-election-two-coalitions-moving-further-apart/.

Phillips, Kevin. *The Emerging Republican Majority*. Princeton, NJ: Princeton University Press, 2014.

Pitts-Taylor, Victoria. *The Brain's Body: Neuroscience and Corporeal Politics*. Durham, NC: Duke University Press, 2016.

Platt, Michael L., and Scott A. Huettel. "Risky Business: The Neuroeconomics of Decision Making under Uncertainty." *Nature Neuroscience* 11, no. 4 (2008): 398–403.

Plato. *Republic*. Edited and translated by Chris Emlyn-Jones and William Preddy. 2 vols. Loeb Classical Library 237 and 276. Cambridge, MA: Harvard University Press, 2013.

———. *Statesman. Philebus. Ion.* Translated by Harold North Fowler and W. R. M. Lamb. Loeb Classical Library 164. Cambridge, MA: Harvard University Press, 1925.

Poulakos, John. "Toward a Sophistic Definition of Rhetoric." *Philosophy & Rhetoric* 16, no. 1 (1983): 35–48.

Powell, Malea. "Rhetorical Powwows: What American Indian Making Can Teach Us about Histories of Rhetorics." Paper presented at Purdue University, Hutton Lecture Series, Lafayette, IN, 2010. http://www.joycerain.com/uploads/2/3/2/0/23207256/_rhetorical_powwows.pdf.

Priestly, Joseph. *Disquisitions Relating to Matter and Spirit.* London, 1777. https://archive.org/details/disquisitionsreloprieuoft.

Prior, Paul, and Jody Shipka. "Chronotopic Lamination: Tracing the Contours of Literate Activity." In *Writing Selves, Writing Societies,* edited by Charles Bazerman and David R. Russell. Fort Collins, CO: WAC Clearinghouse, 2003.

Pruchnic, Jeff. "Neurorhetorics: Cybernetics, Psychotropics, and the Materiality of Persuasion." *Configurations* 16, no. 2 (2008): 167–97.

Puglionesi, Alicia. "The Seductive Allure of Neuroskepticism." *Motherboard,* January 8,2013. https://motherboard.vice.com/en_us/article/788j49/the-seductive-allure-of-neuroskepticism.

Quadflieg, Susanne, David J. Turk, Gordon D. Waiter, Jason P. Mitchell, Adrianna C. Jenkins, and C. Neil Macrae. "Exploring the Neural Correlates of Social Stereotyping." *Journal of Cognitive Neuroscience* 21, no. 8 (2009): 1560–70.

Queen, Mary. "Transnational Feminist Rhetorics in a Digital World." *College English* 70, no. 5 (2008): 471–89.

Racine, Eric, Ofek Bar-Ilan, and Judy Illes. "fMRI in the Public Eye." *Nature Reviews Neuroscience* 6, no. 2 (2005): 159–64.

Ramsay, Ian S., Marco C. Yzer, Monica Luciana, Kathleen D. Vohs, and Angus W. MacDonald. "Affective and Executive Network Processing Associated with Persuasive Antidrug Messages." *Journal of Cognitive Neuroscience* 25, no. 7 (2013): 1136–47.

Ratcliffe, Krista. *Rhetorical Listening: Identification, Gender, Whiteness.* Carbondale: Southern Illinois University Press, 2005.

Reagan, Ronald. "Radio Address to the Nation on Welfare Reform (February 15, 1986)." *The American Presidency Project.* http://www.presidency.ucsb.edu/ws/?pid=36875.

Reid, Thomas. *Essays on the Intellectual Powers of Man.* In *The Works of Thomas Reid,* edited by Sir William Hamilton, 1:213–511. Edinburgh, 1863.

———. *An Inquiry into the Human Mind: On the Principles of Common Sense.* London, 1823.

Rhodes, Rebecca E., Fernando Rodriguez, and Priti Shah. "Explaining the Alluring Influence of Neuroscience Information on Scientific Reasoning." *Journal of Experimental Psychology: Learning, Memory, and Cognition* 40, no. 5 (2014): 1432–40.

Rice, Jenny. *Distant Publics: Development Rhetoric and the Subject of Crisis.* Pittsburgh, PA: University of Pittsburgh Press, 2012.

———. "The New 'New': Making a Case for Critical Affect Studies." *Quarterly Journal of Speech* 94, no. 2 (2008): 200–12.

Richardson, Elaine B. *Hiphop Literacies.* London: Routledge, 2006.

Rickert, Thomas J. *Ambient Rhetoric: The Attunements of Rhetorical Being.* Pittsburgh, PA: University of Pittsburgh Press, 2013.

Rivers, Nathaniel A. "Future Convergences: Technical Communication Research as Cognitive Science." *Technical Communication Quarterly* 20, no. 4 (2011): 412–42.

Rocca, J. *Galen on the Brain: Anatomical Knowledge and Physiological Speculation in the Second Century AD.* Leiden: Brill, 2003.

Rose, F. Clifford. "Cerebral Localization in Antiquity." *Journal of the History of the Neurosciences* 18, no. 3 (2009): 239–47.

Rose, Mike. "Narrowing the Mind and Page: Remedial Writers and Cognitive Reductionism." *College Composition and Communication* 39, no. 3 (1988): 267–302.

Rose, Nikolas S. *Inventing Our Selves: Psychology, Power, and Personhood.* Cambridge: Cambridge University Press, 1998.

Rose, Nikolas S., and Joelle M. Abi-Rached. *Neuro: The New Brain Sciences and the Management of the Mind.* Princeton University Press, 2013.

Royster, Jacqueline Jones, and Gesa Kirsch. *Feminist Rhetorical Practices: New Horizons for Rhetoric, Composition, and Literacy Studies.* Carbondale: Southern Illinois University Press, 2012.

Rudman, Laurie A., and Julie E. Phelan. "The Effect of Priming Gender Roles on Women's Implicit Gender Beliefs and Career Aspirations." *Social Psychology* 41, no. 3 (2010): 192–202.

Rule, Hannah J. "Writing's Rooms." *College Composition and Communication* 69, no. 3 (2018): 402–33.

Saraswati, L. Ayu. "Wikisexuality: Rethinking Sexuality in Cyberspace." *Sexualities* 16, no. 5–6 (2013): 587–603.

Sarbin, Theodore R. "Anxiety: Reification of a Metaphor." *Archives of General Psychiatry* 10, no. 6 (1964): 630–38.

Satel, Sally L., and Scott O. Lilienfeld. *Brainwashed: The Seductive Appeal of Mindless Neuroscience.* New York: Basic Books, 2013.

Savage, Richard. *The Poems of Richard Savage.* Chiswick, 1822.

Schiappa, Edward. *Defining Reality: Definitions and the Politics of Meaning.* Carbondale: Southern Illinois University Press, 2003.

Schreiber, Darren, Greg Fonzo, Alan N. Simmons, Christopher T. Dawes, Taru Flagan, James H. Fowler, and Martin P. Paulus. "Red Brain, Blue Brain: Evaluative Processes Differ in Democrats and Republicans." *PLoS ONE* 8, no. 2 (2013): e52970.

Schulte-Rüther, Martin, Hans J. Markowitsch, N. Jon Shah, Gereon R. Fink, and Martina Piefke. "Gender Differences in Brain Networks Supporting Empathy." *NeuroImage* 42, no. 1 (2008): 393–403.

Schweitzer, N. J., D. A. Baker, and Evan F. Risko. "Fooled by the Brain: Re-examining the Influence of Neuroimages." *Cognition* 129, no. 3 (2013): 501–11.

Scriven, Michael. "The Mechanical Concept of Mind." *Mind* 62, no. 246 (1953): 230–40.

Scurich, Nicholas, and Adam Shniderman. "The Selective Allure of Neuroscientific Explanations." *PLoS ONE* 9, no. 9 (2014): e107529.

Seurinck, R., G. Vingerhoets, F. P. de Lange, and E. Achten. "Does Egocentric Mental Rotation Elicit Sex Differences?" *NeuroImage* 23, no. 4 (2004): 1440–49.

Shah, Carolin, Katharina Erhard, Hanns-Josef Ortheil, Evangelia Kaza, Christof Kessler, and Martin Lotze. "Neural Correlates of Creative Writing: An fMRI Study." *Human Brain Mapping* 34, no. 5 (2013): 1088–101.

Shapin, Steven. "The Politics of Observation: Cerebral Anatomy and Social Interests in the Edinburgh Phrenology Disputes." *Sociological Review* 27 (1979): 139–78.

Shea, Elizabeth Parthenia. *How the Gene Got Its Groove: Figurative Language, Science, and the Rhetoric of the Real.* State University of New York Press, 2008.

Silber, Sherman J. *The Male: From Infancy to Old Age.* New York: Scribner, 1981.

Simpson, Donald. "Phrenology and the Neurosciences: Contributions of F. J. Gall and J. G. Spurzheim." *ANZ Journal of Surgery* 75, no. 6 (2005): 475–82.

Sloop, John M. *Disciplining Gender: Rhetorics of Sex Identity in Contemporary U.S. Culture.* Amherst: University of Massachusetts Press, 2004.

Smith, C. U. M. "The Triune Brain in Antiquity: Plato, Aristotle, Erasistratus." *Journal of the History of the Neurosciences* 19, no. 1 (2010): 1–14.

Smith, Kevin B., Douglas Oxley, Matthew V. Hibbing, John R. Alford, and John R. Hibbing. "Disgust Sensitivity and the Neurophysiology of Left-Right Political Orientations." *PLoS ONE* 6, no. 10 (2011): e25552.

Society for Neuroscience. "The Creation of Neuroscience: The Society for Neuroscience and the Quest for Disciplinary Unity, 1969–1995." *Society for Neuroscience,* 2008. https://web.sfn.org/sfn/about/history-of-sfn/the-creation-of-neuroscience/introduction.

Sontag, Susan. *Illness as Metaphor; and, AIDS and Its Metaphors.* New York: Picador, 2001.

Spence, Janet T. "Gender-Related Traits and Gender Ideology: Evidence for a Multifactorial Theory." *Journal of Personality and Social Psychology* 64, no. 4 (1993): 624–35.

Spencer, Herbert. *The Principles of Psychology.* 3rd ed. 2 vols. London, 1890.

Sprat, Thomas. *The History of the Royal-Society of London.* London, 1667.

Spurzheim, G. *Outlines of Phrenology.* 3rd ed. Boston, 1834.

Stafford, Maria Royne, Nancy E. Spears, and Chung-Kue Hsu. "Celebrity Images in Magazine Advertisements: An Application of the Visual Rhetoric Model." *Journal of Current Issues & Research in Advertising* 25, no. 2 (2003): 13–20.

Stein, Morris I. "Creativity and Culture." *Journal of Psychology* 36, no. 2 (1953): 311–22.

Stormer, Nathan. *Sign of Pathology: U.S. Medical Rhetoric on Abortion, 1800s–1960s.* University Park: Pennsylvania State University Press, 2015.

Stone, Zara. "The Scientists Who Control Your Brain's 'Buy' Button." *OZY,* 2016. http://www.ozy.com/fast-forward/the-scientists-who-control-your-brains-buy-button/70634.

Struthers, William M. *Wired for Intimacy: How Pornography Hijacks the Male Brain.* Downers Grove, IL: IVP Books, 2009.

Tashani, Osama A., Oras A. Alabas, and Mark I. Johnson. "Understanding the Gender–Pain Gap." *Pain Management* 2, no. 4 (2012): 315–17.

Terman, Lewis. *The Measurement of Intelligence: An Explanation of and a Complete Guide for the Use of the Stanford Revision and Extension of the Binet-Simon Intelligence Scale.* Boston: Houghton Mifflin, 1916.

Terrill, Robert. "Mimesis, Duality, and Rhetorical Education." *Rhetoric Society Quarterly* 41, no. 4 (2011): 295–315.

Tesink, Cathelijne M. J. Y., Karl Magnus Petersson, Jos J. A. van Berkum, Daniëlle van den Brink, Jan K. Buitelaar, and Peter Hagoort. "Unification of Speaker and Meaning in Language Comprehension: An fMRI Study." *Journal of Cognitive Neuroscience* 21, no. 11 (2009): 2085–99.

Teston, Christa. "Enthymematic Elasticity in the Biomedical Backstage." In *Topologies as Techniques for a Post-Critical Rhetoric,* edited by Lynda Walsh and Casey Andrew Boyle, 219–42. Cham, Switzerland: Palgrave Macmillan, 2017.

———. "Rendering and Reifying Brain Sex." In *Rhetoric, Through Everyday Things,* edited by Scot Barnet and Casey Boyle, 42–54. Tuscaloosa: University of Alabama, 2016.

Tharp, Twyla, and Mark Reiter. *The Creative Habit: Learn It and Use It for Life; A Practical Guide.* New York: Simon & Schuster, 2006.

Thorndike, Edward L. *The Elements of Psychology.* 2nd ed. New York: A. G. Seiler, 1913.

Thornton, Davi Johnson. *Brain Culture: Neuroscience and Popular Media.* New Brunswick, NJ: Rutgers University Press, 2011.

Tomlinson, Barbara. "To Tell the Truth and Not Get Trapped: Desire, Distance, and Intersectionality at the Scene of Argument." *Signs* 38, no. 4 (2013): 993–1017.

Torrance, E. Paul. "Current Research on the Nature of Creative Talent." *Journal of Counseling Psychology* 6, no. 4 (1959): 309–16.

———. "Scientific Views of Creativity and Factors Affecting Its Growth." *Daedalus* 94: 663–81.

"Tragically Hip's Gord Downie Calls Out to Trudeau During Tour Finale." CBC. 2016. https://news/Canada/tragically-hip-s-gord-downie-calls-out-to-Trudeau-during-tour-finale-1.3729651.

"The Tragically Hip's Last Song Rings Out across Canada." *MacLean's.* 2017. https://www.macleans.ca/videos/video-the-tragically-hips-last-song-rings-out-across-canada/

Trump, Donald. "Our Country Needs a Truly Great Leader." *Wall Street Journal,* June 16, 2015. https://blogs.wsj.com/washwire/2015/06/16/donald-trump-transcript-our-country-needs-a-truly-great-leader/.

Turing, Alan Mathison, and B. Jack Copeland. *The Essential Turing: Seminal Writings in Computing, Logic, Philosophy, Artificial Intelligence, and Artificial Life, Plus the Secrets of Enigma.* Oxford: Clarendon Press, 2004.

Uttal, William R. *The New Phrenology: The Limits of Localizing Cognitive Processes in the Brain.* Cambridge, MA: MIT Press, 2001.

van Dijk, Jelle, Roel Kerkhofs, Iris van Rooij, and Pim Haselager. "Can There Be Such a Thing as Embodied Embedded Cognitive Neuroscience?" *Theory & Psychology* 18, no. 3 (2008): 297–316.

Vidal, Catherine. "Brain, Sex, and Ideology." *Diogenes* 208 (2005): 127–33.

Vine, Betty. "The Neuroscience of Politics." *Brain World,* April 28, 2015. http://brainworldmagazine.com/the-neuroscience-of-politics/.

Wald, Kenneth D., Dennis E. Owen, and Samuel S. Hill. "Political Cohesion in Churches." *Journal of Politics* 52, no. 1 (1990): 197–215.

Walker, Jeffrey. "Of Brains and Rhetorics." *College English* 52, no. 3 (1990): 301–22.

Walter, W. Grey. "An Imitation of Life." *Scientific American* 182, no. 5 (1950): 42–45.

Ward, Adolphus William. *A History of English Dramatic Literature to the Death of Queen Anne.* London, 1875.

Weisberg, Deena Skolnick, Frank C. Keil, Joshua Goodstein, Elizabeth Rawson, and Jeremy R. Gray. "The Seductive Allure of Neuroscience Explanations." *Journal of Cognitive Neuroscience* 20, no. 3 (2008): 470–77.

Weissmann, Jordan. "Donald Trump Explains His Ridiculous Plan to Make Mexico Pay for a Border Fence." *Slate,* August 16, 2015. http://www.slate.com/blogs/moneybox/2015/08/16/donald_trump_on_immigration_build_border_fence_make_mexico_pay_for_it.html.

Weng, Helen Y., Andrew S. Fox, Alexander J. Shackman, Diane E. Stodola, Jessica Z. K. Caldwell, Matthew C. Olson, Gregory M. Rogers, and Richard J. Davidson. "Compassion Training Alters Altruism and Neural Responses to Suffering." *Psychological Science* 24, no. 7 (2013): 1171–80.

Westen, Drew, Pavel S. Blagov, Keith Harenski, Clint Kilts, and Stephan Hamann. "Neural Bases of Motivated Reasoning: An fMRI Study of Emotional Constraints on Partisan Political Judgment in the 2004 U.S. Presidential Election." *Journal of Cognitive Neuroscience* 18, no. 11 (2006): 1947–58.

Wetherell, Margaret. "Trends in the Turn to Affect." *Body & Society* 21, no. 2 (2015): 139–66.

Wiener, Norbert. *Cybernetics; or, Control and Communication in the Animal and the Machine.* New York: John Wiley, 1949.

Williams, Leanne M., Matthew J. Barton, Andrew H. Kemp, Belinda J. Liddell, Anthony Peduto, Evian Gordon, and Richard A. Bryant. "Distinct Amygdala–Autonomic Arousal Profiles in Response to Fear Signals in Healthy Males and Females." *NeuroImage* 28, no. 3 (2005): 618–26.

Williamson, Thad. "Sprawl, Spatial Location, and Politics: How Ideological Identification Tracks the Built Environment." *American Politics Research* 36, no. 6 (2008): 903–33.

Willis, Thomas. *Cerebri Anatome: Cui Accessit Nervorum Descriptio et Usus.* London: Martyn & Allestry, 1664. https://archive.org/details/cerebrianatomecuoowill.

Wilson, Glenn D., and John R. Patterson. "A New Measure of Conservatism." *British Journal of Social and Clinical Psychology* 7, no. 4 (1968): 264–69.

Winterowd, W. Ross. "Style: A Matter of Manner." *Quarterly Journal of Speech* 56, no. 2 (1970): 161–67.

Wolbrecht, Christina. *The Politics of Women's Rights: Parties, Positions, and Change.* Princeton, NJ: Princeton University Press, 2000.

Wolf, Naomi. *Vagina: A Cultural History.* New York: Harper Collins, 2012.

Young, Robert M. *Mind, Brain, and Adaptation in the Nineteenth Century: Cerebral Localization in Its Biological Context from Gall to Ferrier.* Oxford: Clarendon Press, 1970.

Zamboni, Giovanna, Marta Gozzi, Frank Krueger, Jean-René Duhamel, Angela Sirigu, and Jordan Grafman. "Individualism, Conservatism, and Radicalism as Criteria for Processing Political Beliefs: A Parametric FMRI Study." *Social Neuroscience* 4, no. 5 (2009): 367–83.

INDEX

NEW DIRECTIONS IN RHETORIC AND MATERIALITY

BARBARA A. BIESECKER, WENDY S. HESFORD,
AND CHRISTA TESTON, SERIES EDITORS

Current conversations about rhetoric signal a new attentiveness to and critical appraisal of material-discursive phenomena. New Directions in Rhetoric and Materiality provides a forum for responding to and extending such conversations. The series publishes monographs that pair rhetorical theory with an analysis of material conditions and the social-symbolic labor circulating therein. Books in the series offer a "new direction" for exploring the everyday, material, lived conditions of human, nonhuman, and extra-human life—advancing theories around rhetoric's relationship to materiality.

www.ingramcontent.com/pod-product-compliance
Lightning Source LLC
Chambersburg PA
CBHW020350270326
41926CB00007B/377